Linux for Embedded and Real-time Applications

Linux for Embedded and Real-time Applications

by Doug Abbott

Newnes
An Imprint of Elsevier

Amsterdam   Boston   Heidelberg   London   New York   Oxford   Paris   San Diego
San Francisco   Singapore   Sydney   Tokyo

Newnes is an imprint of Elsevier

Copyright © 2003, Elsevier   (USA). All rights reserved.

∞ Recognizing the importance of preserving what has been written,
Elsevier   prints its books on acid-free paper whenever possible.

Library of Congress Cataloging-in-Publication Data

ISBN: 0-7506-7546-2

British Library Cataloguing-in-Publication Data
A catalogue record for this book is available from the British Library.

The publisher offers special discounts on bulk orders of this book.
For information, please contact:

Manager of Special Sales
Elsevier Science
200 Wheeler Road
Burlington, MA 01803
Tel: 781-313-4700
Fax: 781-313-4880

For information on all Newnes publications available, contact our World Wide
Web home page at: http://www.newnespress.com

10 9 8 7 6 5 4 3

Printed in the United States of America

Contents

Preface

" 'You are in a maze of twisty little passages, all alike'

Before you looms one of the most complex and utterly intimidating systems ever written. Linux, the free UNIX clone for the personal computer, produced by a mishmash team of UNIX gurus, hackers, and the occasional loon. The system itself reflects this complex heritage, and although the development of Linux may appear to be a disorganized volunteer effort, the system is powerful, fast, and free. It is a true 32-bit operating system solution."[1]

I have a confession to make. Until about three years ago, I didn't like Unix and avoided it as much as possible. I always considered it deliberately obscure and difficult to use. I still do. Working with Linux has been one of the most frustrating experiences in my long career as a computer engineer. I can do a Windows installation in about 15 minutes without ever referring to a manual. I can't do that with Linux.

But, while Linux is far from being ready for prime time in the world of consumer computing, there are some good things about it that have forced me to soften my bias and grin and bear it. In the embedded space where I work, Linux can no longer be ignored or avoided, nor should it be.

Linux is indeed complex and, unless you're already a Unix guru, the learning curve is quite steep. The information is out there on the web but it is often neither easy to find nor readable. There are probably hundreds of books in print on Linux covering every aspect from beginners' guides to the internal workings of the kernel. But until recently little has been written about Linux in embedded or real-time environments.

---

[1] *Linux Installation and Getting Started*, Matt Welsh, et al.

I decided to climb the Linux learning curve partly because I saw it as an emerging market opportunity and partly because I was intrigued by the Open Source development model. The idea of programmers all over the world contributing to the development of a highly sophisticated operating system just for the fun of it is truly mind-boggling. Having the complete source code not only allows you to modify it to your heart's content, it allows you (in principle at least) to understand how the code works. Unfortunately, my experience has been that a lot of Linux code is "write-only." Someone obviously wrote it, but no one else can read it.

Open Source has the potential to be a major paradigm shift in how our society conducts business because it demonstrates that cooperation can be as useful in developing solutions to problems as competition. Yet at the time this book is being written, serious questions are being raised concerning whether or not it is possible to actually make money with Open Source software. Is there a business model that works? The jury is still out.

Audience and Prerequisites

This book is directed at two different audiences:

- The primary audience is embedded programmers who need an introduction to Linux in the embedded space. This is where I came from and how I got into Linux so it seems like a reasonable way to structure the book.

- The other audience is Linux programmers who need an introduction to the concepts of embedded and real-time programming.

Consequently, each group will see some material that is review although it may be presented with a fresh perspective.

This book is not a beginners' guide. I assume that you have successfully installed a Linux system and have at least played around with it some. You know how to log in, you've experimented with some of the command utilities and have probably fired up X-windows. Chapter 2 is a cursory introduction to some of the features and characteristics of Linux that are of interest to embedded and real-time programmers.

The book is divided into two parts: Part I deals with Linux in the embedded space, Part II looks at different approaches to giving the Linux kernel deterministic characteristics. It goes without saying that you can't learn to program by reading a book. You have to do it. That's why this book is designed as a practical hands-on guide. The companion CD contains several packages that we'll explore in depth in the following chapters.

Embedded programming implies a target machine that is separate and distinct from the workstation development machine. We'll look at two target environments developing essentially the same projects on each.

- The X86. More and more embedded projects are choosing the PC architecture as a target platform. And of course Linux was originally developed for the PC. For the purpose of this book, an x86 target need be nothing more than an old 486 box gathering dust in your closet.

- Motorola Coldfire. The Coldfire processor is a variant on the 68000. From the viewpoint of Linux, the interesting thing about the Coldfire is that it lacks a memory management unit. This necessitates a modification to the kernel to fake out the memory protection mechanism. A suitable Coldfire target board is available for a reasonable price from Lineo, Inc. This is an example of a small processor suitable for deeply embedded applications.

Personal Biases

Like most computer users, for better or worse, I've spent years in front of a Windows screen. But before that I was perfectly at home with DOS and even before that I hacked away at RT-11, RSX-11 and VMS. So it's not like I don't understand command line programming. In fact it was probably a couple of years before I finally added WIN to my AUTOEXEC.BAT file.

Hardcore Unix programmers think GUIs are for wimps. They proudly do everything from the command line. Say what you will, but I like GUIs. Yes, the command line still has its place, particularly for shell scripts and makefiles, but for moving around the file hierarchy and doing simple file operations like move, copy, delete, rename, etc, drag-and-drop beats the miserably obscure Unix commands hands down. I also refuse to touch text-based editors like vi and emacs. Sure they're powerful if you can remember all the obscure commands. Give me a WYSIWYG editor any day.

My favorite GUI is the KDE desktop environment. It has all the necessary bells and whistles including a very nice syntax coloring editor. KDE is included in most commercial Linux distributions. Clearly, you're free to use whatever environment you're most comfortable with to work the book's examples. But if you're new to Linux, I would recommend KDE.

OK, enough philosophizing. Let's get on with it. Join me for a thrill-packed, sometimes bumpy, but ultimately fun and rewarding, ride through those twisty little passages known as Linux.

## CHAPTER 1

The Embedded and Real-time Space

What Is Embedded?

You're at a party when an attractive member of the opposite sex approaches and asks you what you do. You could be flip and say something like "as little as possible," but eventually the conversation will get around to the fact that you write software for embedded systems. Before your new acquaintance starts scanning the room for a lawyer or doctor to talk to, you'd better come up with a captivating explanation of embedded systems.

I usually start by saying that an embedded system is a device that has a computer inside it, but the user of the device doesn't necessarily know, or care, that the computer is there. It's hidden. The example I usually give is the engine control computer in your car. You don't drive the car any differently because the engine happens to be controlled by a computer. In fact the typical car today has more raw computing power than the Lunar Lander. You can then go on to point out that there are a lot more embedded computers out in the world than there are PCs—by at least an order of magnitude. The average house contains perhaps a couple dozen computers even if it doesn't have a PC.

From the viewpoint of programming, embedded systems show a number of significant differences from conventional "desktop" applications. For example, most desktop applications deal with a fairly predictable set of I/O devices—a disk, graphic display, a keyboard, mouse, sound card, perhaps a network interface. And these devices are generally well supported by the operating system. The application programmer doesn't need to pay much attention to them.

Embedded systems on the other hand often incorporate a wider variety of input/output devices than typical desktop computers. A system may include user I/O in the form of switches, pushbuttons and various types of displays. It may have one or more communication channels, either asynchronous serial or network ports. It may implement data acquisition and control in the form of analog-to-digital (A/D) and digital-to-analog (D/A) converters. These devices seldom have the kind of operating system support that application programmers are accustomed to. The embedded systems programmer often has to deal directly with the hardware.

What Is Real-time?

Real-time is harder to explain. The basic idea behind real-time is that we expect the computer to respond to its environment *in time*. Many people assume that real-time means real fast. Not true. Real-time simply means *fast enough* in the context in which the system is operating. If we're talking about the computer that runs your car's engine, that's fast! That guy has to make decisions—about fuel flow, spark timing—every time the engine makes a revolution.

On the other hand, consider a chemical refinery controlled by one or more computers. The computer system is responsible for controlling the process and detecting potentially destructive malfunctions. But chemical processes have a time constant in the range of seconds to minutes at the very least. So we would assume that the computer system should be able to respond to any malfunction in sufficient time to avoid a catastrophe. But suppose the computer were in the midst of printing an extensive report about last week's production when the malfunction occurred. How soon would it be able to respond to the potential emergency?

The essence of real-time computing is not only that the computer responds to its environment fast enough, but that it responds *reliably* fast enough. The engine control computer must be able to adjust fuel flow and spark timing every time the engine turns over. If it's late, the engine doesn't perform right. The controller of a chemical plant must be able to detect and respond to abnormal conditions in sufficient time to avoid a catastrophe. If it doesn't, it has failed.

So the art of real-time programming is designing systems that reliably meet timing constraints in the midst of random asynchronous events. Not surprisingly, this is easier said than done and there is an extensive body of literature and development work devoted to the theory of real-time systems.

How and Why Does Linux Fit in?

By now just about everyone in the computer business knows the history of Linux: how Linus Torvalds started it all back in 1991 as a simple hobby project to which he invited other interested hackers to contribute. Back then no one could have predicted that this amorphous consortium of volunteer programmers and the occasional loon, connected only by the Internet, would produce a credible operating system to compete with even the Borg of Redmond.

Of course, Linux developed as a general-purpose operating system in the model of Unix whose basic architecture it emulates. No one would suggest that Unix is suitable as an embedded operating system. It's big, it's a resource hog and its scheduler is based on "fairness" rather than priority. In short, it's the exact antithesis of an embedded operating system.

But Linux has several things going for it that earlier versions of Unix lack. It's free and you get the source code. There is a large and enthusiastic community of Linux developers and users. There's a good chance that someone else either is working or has worked on the same problem you're facing. It's all out there on the web. The trick is finding it.

Open Source

Linux has been developed under the philosophy of Open Source software pioneered by the Free Software Foundation. Open Source is based on the notion that software should be freely available: to use, to modify, to copy.

There are a number of misconceptions about the nature of Open Source software. Perhaps the best way to explain what it is is to start by talking about what it isn't.

- Open Source is not shareware. A precondition for the use of shareware is that you pay the copyright holder a fee. Open source code is freely available and there is no obligation to pay for it.

- Open Source is not "public domain." Public domain code, by definition, is not copyrighted. Open Source code is copyrighted by its author who has released it under the terms of the GNU General Public License (GPL). The copyright owner thus gives you the right to use the code provided you adhere to the terms of the GPL.

- Open Source is not necessarily free of charge. Having said that there's no obligation to pay for Open Source software doesn't preclude you from charging a fee to package and distribute it. A number of companies are in the specific business of selling packaged "distributions" of Linux.

Why would you pay someone for something you can get for free? Presumably because everything is in one place and you can get some support from the vendor. Of course the quality of support greatly depends on the vendor.

So "free" refers to freedom to use the code and not necessarily zero cost. Think "free speech," not "free beer."

Open Source code is:

- Subject to the terms of the GNU Public License (see below).

- Subject to critical peer review. As an Open Source programmer, your code is out there for everyone to see and the Open Source community tends to be a very critical group. Open Source code is subject to extensive testing and peer review. It's a Darwinian process in which only the best code survives. "Best" of course is a subjective term. It may be the best *technical* solution but it may also be completely unreadable.

- Highly subversive. The Open Source movement subverts the dominant paradigm, which says that intellectual property such as software must be jealously guarded so you can make a lot of money off of it. In contrast, the Open Source philosophy is that software should be freely available to everyone for the maximum benefit of society. Richard Stallman, founder of the Free Software Foundation, is particularly vocal in advocating that software should not have owners (see Appendix C).

[handwritten annotation: contradicting to established view]

Not surprisingly, Microsoft sees Open Source as a serious threat to its business model. Microsoft representatives have gone so far as to characterize Open Source as "un-American." On the other hand, many leading vendors of Open Source software give their programmers and engineers company time to contribute to the Open Source community. And it's not just charity, it's good business.

The GNU Public License (GPL)

Open Source software is released according to the terms of the GNU Public License, GPL. Unlike most End User License Agreements (EULA) for software, whose motivation is to restrict your rights, the GPL is intended to guarantee your rights to use, modify and copy the subject software. Along with the rights comes an obligation. If you modify and subsequently distribute software covered by the GPL, you are obligated to make available the modified source code. The changes become a "derivative work" which is also subject to the terms of the GPL. This allows other users to understand the software better and to make further changes if they wish.

This works well for most software but there is at least one problem. Suppose, for example, that you write a clever application that you wish to keep proprietary. If you link it with a C library covered by the GPL, your application becomes a derivative work and thus you're required to distribute your source code.

To get around this, and therefore promote the development of Open Source libraries, the Free Software Foundation came up with the "Library GPL." The distinction is that a program linked to a library covered by the LGPL is not considered a derivative work and so there's no requirement to distribute the source, although you must still distribute the source to the library itself.

Subsequently, the LGPL became known as the "Lesser GPL" because it offers less freedom to the user. So while the LGPL makes it possible to develop proprietary products using Open Source software, the FSF encourages developers to place their libraries under the GPL in the interest of maximizing openness.

Portable and Scalable

Linux was originally developed for the Intel x86 family of processors and most of the ongoing kernel development work continues to be on x86s. Nevertheless, the design of the Linux kernel makes a clear distinction between processor-dependent code that must be modified for each different architecture, and code that can be ported simply by recompiling it. Consequently, Linux has been ported to a wide range of processor architectures including:

- Motorola 68k and its many variants
- Alpha
- Power PC
- ARM
- SPARC
- MIPS

to name just a few. So whatever 32-bit architecture you're considering for your embedded project, chances are there's a Linux port available for it and a community of developers supporting it.

A typical desktop Linux installation runs into several hundred megabytes of disk space and requires 32 megabytes of RAM to execute decently. By contrast, embedded targets are often limited to perhaps eight or 16 megabytes of RAM and a few megabytes of ROM or flash. Fortunately, Linux is highly modular. Much of that several hundred megabytes represents documentation, desktop utilities and options like games that simply aren't necessary in an embedded target. It is not difficult to produce a fully functional, if limited, Linux system occupying no more than 2 megabytes of flash memory.

The kernel itself is highly configurable and includes reasonably user-friendly tools that allow you to remove kernel functionality not required in your application.

Resources

Linux resources on the web are extensive. This is a list of some sites that are of particular interest to embedded developers.

linuxdevices.com – A news and portal site devoted to the entire range of issues surrounding embedded Linux.

embedded-linux.org – The Embedded Linux Consortium, a nonprofit, vendor-neutral trade association promoting Linux in the embedded space. Its major effort at present is the development of an embedded Linux platform specification.

kernel.org – The Linux kernel archive. This is where you can download the latest kernel versions as well as virtually any previous version.

sourceforge.net – "World's largest Open Source development website." Provides free services to open source developers including project hosting and management, version control, bug and issue tracking, backups and archives, and communication and collaboration resources.

embedded.com – The web site for *Embedded Systems Programming* magazine. This site is not specifically oriented to Linux, but is quite useful as a more general embedded information tool.

CHAPTER 2

Introducing Linux

Software is like sex, it's better when it's free.
— Linus Torvalds

For those who may be new to Unix-style operating systems, this chapter provides an introduction to some of the salient features of Linux, especially those of interest to embedded developers. This is by no means a thorough introduction and there are many books available that delve into these topics in much greater detail.

Feel free to skim, or skip this chapter entirely, if you are already comfortable with Unix concepts.

Features

Here are some of the important features of Linux, and Unix-style operating systems in general.

- *Multitasking* – The Linux scheduler implements true, preemptive multitasking in the sense that a higher priority process made ready by the occurrence of an asynchronous event will preempt the currently running process. However, the stock Linux kernel itself is not preemptible[1]. So a process may not be preempted while it is executing a kernel service. Some kernel services can be rather long and the resulting latencies make standard Linux generally unsuitable for real-time applications.

---

[1] Is it "preemptible" or "preemptable"? Word's spelling checker says they're both wrong. A debate on linuxdevices.com a while back seemed to come down on the side of "ible" but not conclusively. I think I'll stick with preemptible.

- *Multi-user* – Unix evolved as a time-sharing system that allowed multiple users to share an expensive (at that time anyway) computer. Thus there are a number of features that support privacy and data protection. Linux preserves this heritage and puts it to good use in server environments.[2]

- *Multi-processing* – Linux offers extensive support for true symmetric multi-processing (SMP).

- *Protected Memory* – Each Linux process operates in its own private memory space and is not allowed to directly access the memory space of another process. This prevents a wild pointer in one process from damaging the memory space of another process. The errant access is trapped by the processor's memory protection hardware and the process is terminated with appropriate notification.

- *Hierarchical File System* – Yes, all modern operating systems—even DOS—have hierarchical file systems. But the Linux/Unix model adds a couple of nice wrinkles on top of what we're used to with traditional PC operating systems:

    o *Links* – A link is simply a file system entry that points to another file rather than being a file itself. Links can be a useful way to share files among multiple users and find extensive use in configuration scenarios for selecting one of several optional files.

    o *Device-Independent I/O* – Again, this is nothing new, but Linux takes the concept to its logical conclusion by treating every peripheral device as an entry in the file system. From an application's viewpoint, there is absolutely no difference between writing to a file and writing to, say, a printer.

[handwritten: links are like Windows shortcuts]

---

[2] Although my experience in the embedded space is that the protection features, particularly file permissions, can be a downright nuisance.

Protected Mode Architecture

The implementation of protected mode memory in contemporary Intel processors first made its appearance in the 80386. It utilizes a full 32-bit address for an addressable range of 4 gigabytes. Access is controlled such that a block of memory may be: Executable, Read only or Read/Write.

The processor can operate in one of four *privilege levels*. A program running at the highest privilege level, level 0, can do anything it wants—execute I/O instructions, enable and disable interrupts, modify descriptor tables. Lower privilege levels prevent programs from performing operations that might be "dangerous." A word processing application probably shouldn't be messing with interrupt flags, for example. That's the job of the operating system.

So application code typically runs at the lowest level while the operating system runs at the highest level. Device drivers and other services may run at the intermediate levels. In practice, however, Linux and most other operating systems for Intel processors only use levels 0 and 3. In Linux level 0 is called "Kernel Space" while level 3 is called "User Space."

Real Mode

To begin our discussion of protected mode programming in the x86, it's useful to review how "real" address mode works. Back in the late '70s when Intel was designing the 8086, the designers faced the dilemma of how to access a megabyte of address space with only 16 bits. At the time a megabyte was considered an immense amount of memory. The solution they came up with, for better or worse, builds a 20-bit (1 mega-byte) address out of two 16-bit quantities called the segment and offset. Shifting the segment value four bits to the left and adding it to the offset creates the 20-bit linear address (see figure 2-1).

In real mode, x86 processors have four segment registers. Every reference to memory derives its segment value from one of these registers. By default, instruction execution is relative to the Code Segment (CS). Most data references (MOV for example) are relative to the Data Segment (DS) and instructions that reference the stack are relative to the Stack Segment (SS).

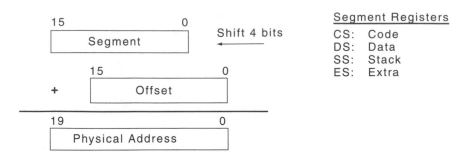

Figure 2-1: X86 Real Mode Addressing

The Extra Segment (ES) is used in string move instructions and can be used whenever an extra data segment is needed. The default segment selection can be overridden with segment prefix instructions.

A segment can be up to 64 kilobytes long and is aligned on 16-byte boundaries. Programs less than 64 kilobytes are inherently position-independent and can be easily relocated anywhere in the 1 megabyte address space. Programs larger than 64 kilobytes, either in code or data, require multiple segments and must explicitly manipulate the segment registers.

Protected Mode

Protected mode still makes use of the segment registers, but instead of providing a piece of the address directly, the value in the segment register (now called the *selector*) becomes an index into a table of *segment descriptors*. The segment descriptor fully describes a block of memory including, among other things, its base and limit (see Figure 2-2). The linear address in physical memory is computed by adding the offset in the logical address to the base contained in the descriptor. If the resulting address is greater than the limit specified in the descriptor, the processor signals a memory protection fault.

A descriptor is an 8-byte object that tells us everything we need to know about a block of memory.

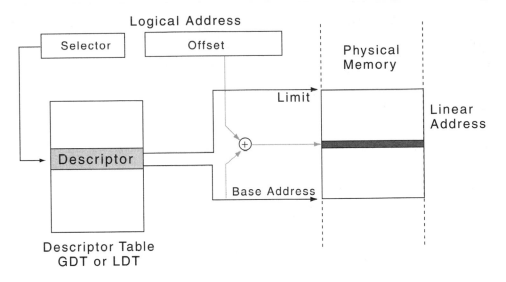

Figure 2-2: Protected Mode Address Calculation

Base Address[31:0] Starting address for this block/segment.

Limit[19:0] Length of this segment. This may be either the length in bytes (up to 1 megabyte) or the length in 4 kilobyte *pages*. The interpretation is defined by the Granularity bit.

Type A 4-bit field that defines the kind of memory that this segment describes

S 0 = This descriptor describes a "System" segment. 1 = This descriptor describes a code or data segment.

DPL Descriptor Privilege Level. A 2-bit field that defines that minimum privilege level required to access this segment.

P Present. 1 = The block of memory represented by this descriptor is present in memory. Used in paging.

G Granularity. 0 = Interpret Limit as bytes. 1 = Interpret Limit as 4 kilobyte pages

Note that with the Granularity bit set to 1, a single segment descriptor can represent the entire 4 gigabyte address space.

Normal descriptors (S bit = 1) describe memory blocks representing data or code. The Type field is four bits where the most significant bit distinguishes between *code* and *data* segments. Code segments are executable, data segments are not. A code segment may or may not also be readable. A data segment may be writable. Any attempted access that falls outside the scope of the Type field—attempting to execute a data segment for example—causes a memory protection fault.

"Flat" vs. Segmented Memory Models

Because a single descriptor can reference the full 4-gigabyte address space, it is possible to build your system by reference to a single descriptor. This is known as "flat" model addressing and is, in effect, a 32-bit equivalent of the addressing model found in most 8-bit microcontrollers as well as the "tiny" memory model of DOS. All memory is equally accessible and there is no protection.

In order to take advantage of the protection features built into Protected Mode, you must allocate different descriptors for the operating system and applications. Figure 2-3 illustrates the difference between flat model and segmented model.

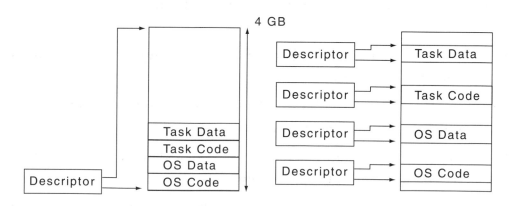

Figure 2-3: "Flat" vs. Segmented Addressing

Paging

Paging is the mechanism that allows each task to pretend that it owns a very large flat address space. That space is then broken down into 4 kilobyte *pages*. Only the pages currently being accessed are kept in main memory. The others reside on disk.

As shown in Figure 2-4, paging adds another level of indirection. The 32-bit linear address derived from the selector and offset is divided into three fields. The high-order ten bits serve as an index into the *Page Directory*. The Page Directory Entry (PDE) points to a *Page Table*. The next ten bits in the linear address provide an index into that table. The Page Table Entry (PTE) provides the base address of a 4 kilobyte page in physical memory called a *Page Frame*. The low-order twelve bits of the original linear address supplies the offset into the page frame. Each task has its own Page Directory pointed to by processor control register CR3.

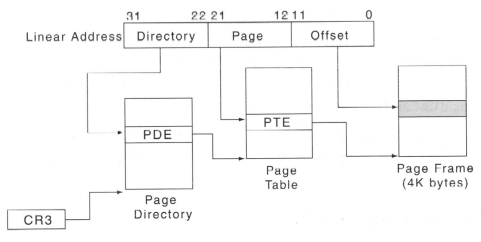

Figure 2-4: Paging

At either stage of this lookup process it may turn out that either the Page Table or the Page Frame is not present in physical memory. This causes a *Page Fault*, which in turn causes the operating system to find the corresponding page on disk and load it into an available page in memory. This in turn may require "swapping out" the page that currently occupies that memory.

A further advantage to paging is that it allows multiple tasks or processes to easily share code and data by simply mapping the appropriate sections of their individual address spaces into the same physical pages.

Paging is optional—you don't have to use it, although Linux does. Paging is controlled by a bit in processor register CR0.

Page Directory and Page Table entries are each four bytes long, so the Page Directory and Page Tables are a maximum of 4 kilobytes, which also happens to be the Page Frame size. The high-order 20 bits point to the base of a Page Table or Page Frame. Bits 9 to 11 are available to the operating system for its own use. Among other things, these could be used to indicate that a page is to be "locked" in memory—i.e., not swappable.

Of the remaining control bits the most interesting are:

P   Present. 1 = this page is in memory. If this bit is 0, referencing this Page Directory or Page Table entry causes a page fault. Note that when P == 0 the remainder of the entry is not relevant.

A   Accessed. 1 = this page has been read or written. Set by the processor but cleared by the OS. By periodically clearing the Accessed bits, the OS can determine which pages haven't been referenced in a long time and are therefore subject to being swapped out.

D   Dirty. 1 = this page has been written. Set by the processor but cleared by the OS. If a page has not been written to, there is no need to write it back to disk when it has to be swapped out.

The Linux Process Model

The basic structural element in Linux is a *process* consisting of executable code and a collection of *resources* like data, file descriptors and so on. These resources are fully protected such that one process can't directly access the resources of another. In order for two processes to communicate with each other, they must use the inter-process communication mechanisms defined by Linux, such as shared memory regions or pipes.

This is all well and good as it establishes a high degree of protection in the system. An errant process will most likely be detected by the system and thrown out before it can do any damage to other processes (see Figure 2-5). But there's a price to be paid in terms of excessive overhead in creating processes and using the inter-process communication mechanisms.

UNIX Process Model

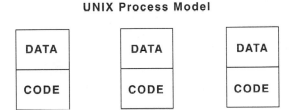

Multi-threaded Process

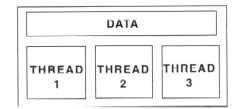

Figure 2-5: "Processes" vs. "Threads"

A *thread* on the other hand is code only. Threads only exist within the context of a process and all threads in one process share its resources. Thus, all threads have equal access to data memory and file descriptors. This model is sometimes called *lightweight multitasking* to distinguish it from the Unix/ Linux process model.

The advantage of lightweight tasking is that inter-thread communication is more efficient. The drawback, of course, is that any thread can clobber any other thread's data. Historically, most real-time operating systems have been structured around the lightweight model. In recent years the cost of memory protection hardware has dropped dramatically. In response, many RTOS vendors now offer protected mode versions of their systems that look like the Linux process model.

The fork() function

Linux starts life with one process, the init process, created at boot time. Every other process in the system is created by invoking fork(). The process calling fork() is termed the *parent* and the newly created process is termed the *child*. So every process has ancestors and may have descendants, depending on who created who.

If you've grown up with multitasking operating systems where tasks are created from functions by calling a task creation service, the fork process can seem downright bizarre. fork() creates a *copy of the parent process*—code, data, file descriptors and any other resources the parent may currently hold. This could add up to megabytes of memory space to be copied. To avoid copying a lot of stuff that may be overwritten anyway, Linux employs a *copy-on-write* strategy.

fork() begins by making a copy of the process data structure and giving it a new process identifier (PID) for the child process. Then it makes a new copy of the Page Directory and Page Tables. Initially the page table entries all point to the same physical pages as the parent process. All pages for both processes are set to read-only. When one of the processes tries to write, that causes a page fault, which in turn causes Linux to allocate a new page for that process and copy over the contents of the existing page.

Listing 2-1: Trivial Example of Fork

```
#include <unistd.h>
#include <>

pid_t pid;

void do_child_thing (void)
{
    printf ("I am the child. My PID is %d\n", pid);
}
void do_parent_thing (void)
{
    printf ("I am the parent. My child's PID is %d\n", pid);
}
```

```
void main (void)
{
    switch (pid = fork())
    {
    case –1:
        printf ("fork failed\n");
        break;
    case 0:
        do_child_thing();
        break;
    default:
        do_parent_thing();
    }
    exit (0);
}
```

Since both processes are executing the same code, they both continue from the return from fork() (this is what's so bizarre!). In order to distinguish parent from child, fork() returns a function value of 0 to the child process but returns the PID of the child to the parent process. Listing 2-1 is a trivial example of the fork call.

clone() is a Linux-specific variation on fork(), the difference being that the former offers greater flexibility in specifying how much of the parent's operating environment is shared with the child. This is the mechanism that is used to implement Posix threads (see Chapter 11) at the kernel level.

The execve() function

Of course, what really happens 99% of the time is that the child process invokes a new program by calling execve() to load an executable image file from disk. Listing 2-2 shows in skeletal form a simple command line interpreter. It reads a line of text from stdin, parses it and calls fork() to create a new process. The child then calls execve() to load a file and execute the command just entered. execve() overwrites the calling process's code, data and stack segments.

If this is a normal "foreground" command, the command interpreter must wait until the command completes. This is accomplished with waitpid() which blocks the calling process until the process matching the pid argument has completed. Note, by the way, that most multitasking operating systems do not have the ability to block one process or task pending the completion of another.

If execve() succeeds, it does not return. Instead, control is transferred to the newly loaded program.

Listing 2-2: Command Line Interpreter

```c
#include <unistd.h>
void main (void)
{
    char *argv[10], *filename;
    char text[80];
    char foreground;
    pid_t pid;
    int status;
    while (1)
    {
        gets (text);
// Parse the command line to derive filename and
// arguments. Decide if it's a foreground command.
        switch (pid = fork())
        {
            case -1:
                printf ("fork failed\n");
                break;
            case 0:  // child process
                if (execve (filename, argv, NULL) < 0)
                    printf ("command failed\n");
                break;
            default: // parent process
                if (foreground)
                    waitpid (&status, pid);
        }
    }
}
```

Try it out

Its interesting to see how many processes Linux spawns just by booting up. Reboot your system, log in and execute:

```
ps -A | more
```

PID	TTY	TIME	CMD		PID	TTY	TIME	CMD
1	?	00:00:04	init		565	?	00:00:00	gpm
2	?	00:00:00	kflushd		600	?	00:00:00	xfs
3	?	00:00:00	kupdate		648	?	00:00:00	rpc.rquotad
4	?	00:00:00	kpiod		657	?	00:00:00	rpc.mountd
5	?	00:00:00	kswapd		666	?	00:00:00	nfsd
6	?	00:00:00	mdrecoveryd		667	?	00:00:00	nfsd
332	?	00:00:00	portmap		668	?	00:00:00	nfsd
347	?	00:00:00	lockd		669	?	00:00:00	nfsd
348	?	00:00:00	rpciod		670	?	00:00:00	nfsd
357	?	00:00:00	rpc.statd		671	?	00:00:00	nfsd
371	?	00:00:00	apmd		672	?	00:00:00	nfsd
422	?	00:00:00	syslogd		673	?	00:00:00	nfsd
431	?	00:00:00	klogd		677	tty1	00:00:00	login
445	?	00:00:00	identd		678	tty2	00:00:00	mingetty
449	?	00:00:00	identd		679	tty3	00:00:00	mingetty
450	?	00:00:00	identd		680	tty4	00:00:00	mingetty
451	?	00:00:00	identd		681	tty5	00:00:00	mingetty
452	?	00:00:00	identd		682	tty6	00:00:00	mingetty
463	?	00:00:00	atd		685	tty1	00:00:00	bash
477	?	00:00:00	crond		709	tty1	00:00:00	ps
506	?	00:00:00	lpd		710	tty1	00:00:00	bash
550	?	00:00:00	sendmail					

—More—

This comes from Red Hat 6.2

The Linux Filesystem

The Linux filesystem is in many ways similar to the filesystem you might find on a Windows PC or a Macintosh. It's a hierarchical system that lets you create any number of subdirectories under a root directory identified by "/". Like Windows, file names can be very long. However in Linux, as in most Unix-like systems, filename "extensions," the part of the filename following ".", have much less meaning. For example, while Windows executables always have the extension ".exe", Linux executables rarely have an extension at all. By and large, the contents of a file are identified by a file header rather than a specific extension identifier.

Unlike Windows, file names in Linux are *case-sensitive*. Therefore, Foobar is a different file from foobar is different from fooBar. Sorting is also case-sensitive. File names beginning with uppercase letters appear before those that begin with lowercase letters in directory listings sorted by name. File names that begin with "." are considered to be "hidden" and are not displayed in directory listings unless you specifically ask for them.

Additionally, the Linux filesystem has a number of features that go beyond what you find in a typical Windows system. Let's take a look at some of the features that may be of interest to embedded programmers.

File Permissions

Because Linux is multi-user, every file has a set of "permissions" associated with it to specify what various classes of users are allowed to do with that file. Get a detailed listing of some Linux directory, either by entering the command ls –l in a console window or with the desktop file manager. Part of the entry for each file is a set of 10 flags and a pair of names that look something like this:

```
-rw-r—r—        Andy    physics
```

In this example, Andy is the "owner" of the file and he belongs to a "group" of users called physics, perhaps the physics department at some university. Generally, but not always, the owner is the person who created the file.

The first of the ten flags identifies the file type. Ordinary files get a dash here. Directories are identified by "d", links are "l" and so on. The remaining nine flags divide into three groups of three flags each. The flags are the same for all groups and represent, respectively, permission to read the file, "r", write the file, "w", or execute the file if it's an executable, "x". Write permission also allows the file to be deleted.

The three groups then represent the permissions granted to different classes of users. The first group identifies the permissions granted the owner of the file and virtually always allows reading and writing. The second flag group gives permissions to other members of the same group of users. In this case the physics group has read access to the file but not write access. The final flag group gives permissions to the "world"—i.e., all users.

The "root" User

There's one very special user, named "root," in every Linux system. Root can do anything to any file regardless of the permission flags. Root is primarily intended for system administration purposes and is not recommended for day-to-day use. Clearly you can get in a lot of trouble if you're not careful and root privileges pose a potential security threat. Nevertheless, the kinds of things that embedded and real-time developers do with the system often require write or executable access to files owned by root and thus require you to be logged in as the root user.

If you're logged on as a normal user, you can switch to being root with the su, substitute user, command. The su command with no arguments starts up a shell with root privileges provided you enter the correct password. To return to normal user status, terminate the shell by typing ^d or exit. Most of the time I just log in as root because it's less hassle and I'm not too worried about hackers compromising my relatively isolated system.

The /proc filesystem

The /proc file system is an interesting feature of Linux. It acts just like an ordinary file system. You can list the files in the /proc directory, you can read and write the files, but they don't really exist. The information in a /proc file is generated on the fly when the file is read. The kernel module that registered a given /proc file contains the functions that generate read data and accept write data.

/proc files are another window into the kernel. They provide dynamic information about the state of the system in a way that is easily accessible to user-level tasks and the shell. In the abbreviated directory listing of Figure 2-6, the directories with number labels represent processes. Each process gets a directory under /proc with several files describing the state of the process.

```
ls -l /proc
total 0
dr-xr-xr-x   3 root     root        0Aug 25 15:23 1
dr-xr-xr-x   3 root     root        0Aug 25 15:23 2
dr-xr-xr-x   3 root     root        0Aug 25 15:23 3
dr-xr-xr-x   3 bin      root        0Aug 25 15:23 303
dr-xr-xr-x   3 nobody   nobody      0Aug 25 15:23 416
dr-xr-xr-x   3 daemon   daemon      0Aug 25 15:23 434
dr-xr-xr-x   3 xfs      xfs         0Aug 25 15:23 636
dr-xr-xr-x   4 root     root        0Aug 25 15:23 bus
-r--r--r--   1 root     root        0Aug 25 15:23 cmdline
-r--r--r--   1 root     root        0Aug 25 15:23 cpuinfo
-r--r--r--   1 root     root        0Aug 25 15:23 devices
-r--r--r--   1 root     root        0Aug 25 15:23 filesystems
dr-xr-xr-x   2 root     root        0Aug 25 15:23 fs
dr-xr-xr-x   4 root     root        0Aug 25 15:23 ide
-r--r--r--   1 root     root        0Aug 25 15:23 interrupts
-r--r--r--   1 root     root        0Aug 25 15:23 ioports
```

Figure 2-6: The /proc Filesystem

Try it out

```
[root@lab /Doug]# cd /proc
[root@lab /proc]# cat interrupts
          CPU0
   0:     555681    XT-PIC  timer
   1:     110       XT-PIC  keyboard
   2:     0         XT-PIC  cascade
   7:     1         XT-PIC  soundblaster
   8:     1         XT-PIC  rtc
   9:     1670      XT-PIC  DC21041 (eth0)
  12:     11342     XT-PIC  PS/2 Mouse
  13:     1         XT-PIC  fpu
  14:     203260    XT-PIC  ide0
 NMI:     0
```

The interrupts proc file tells you what interrupt sources have been registered by what device drivers and how many times each interrupt has triggered. To prove that the file data is being created dynamically, repeat the same command.

```
[root@lab /proc]# cat interrupts
        CPU0
   0:      556540      XT-PIC  timer
   1:         116      XT-PIC  keyboard
   2:           0      XT-PIC  cascade
   7:           1      XT-PIC  soundblaster
   8:           1      XT-PIC  rtc
   9:        1672      XT-PIC  DC21041 (eth0)
  12:       11846      XT-PIC  PS/2 Mouse
  13:           1      XT-PIC  fpu
  14:      203376      XT-PIC  ide0
 NMI:           0
```

Not surprisingly, most of the numbers have gone up.

The Filesystem Hierarchy Standard (FHS)

A Linux system typically contains a very large number of files. For example, a typical Red Hat installation may contain around 30,000 files occupying close to 400 megabytes of disk space. Clearly it's imperative that these files be organized in some consistent, coherent manner. That's the motivation behind the Filesystem Hierarchy Standard, FHS. The standard allows both users and software developers to "predict the location of installed files and directories."[3] FHS is by no means specific to Linux. It applies to Unix-like operating systems in general.

FHS specifies several directories and their contents directly subordinate to root. This is illustrated in Figure 2-7. The FHS starts by characterizing files along two independent axes:

- *Sharable vs. non-sharable.* A networked system may be able to mount certain directories through NFS such that multiple users can share executables. On the other hand, some information is unique to a specific computer and is thus not sharable.

---

[3] *Filesystem Hierarchy Standard — Version 2.2 final,* edited by Rusty Russell and Daniel Quinlan. Available from www.pathname.com/fhs

- *Static vs. variable.* Many of the files in a Linux system are executables that don't change, they're *static*. But the files that users create or acquire, by downloading or e-mail for example, are *variable*. These two classes of files should be cleanly separated.

```
/                      the root directory
    ── bin             Essential command binaries
    ── boot            Static files of the boot loader
    ── dev             Device "inode" files
    ── etc             Host-specific system contiguration
    ── home            Home directories for individual users (optional)
    ── lib             Essential shared libraries and kernel modules
    ── mnt             Mount point for temporarily mounting a filesystem
    ── opt             Additional application software packages
    ── root            Home directory for the root user (optional)
    ── sbin            Essential system binaries
    ── tmp             Temporary files
    ── usr             Secondary hierarchy
    ── var             Variable data
```

Figure 2-7: Filesystem Hierarchy

Here is a description of the directories defined by FHS

- /bin Contains binary executables of commands used both by users and the system administrator. FHS specifies what files /bin must contain. These include among other things the command shell and basic file utilities. /bin files are static and sharable.

- /boot Contains everything required for the boot process except configuration files and the map installer. In addition to the kernel executable image, /boot contains data that is used before the kernel begins executing user-mode programs. /boot files are static and non-sharable.

- /etc Contains host-specific configuration files and directories. With the exception of mtab, which contains dynamic information about filesystems, /etc files are static. FHS identifies three optional subdirectories of /etc:

o   /opt Configuration files for add-on application packages contained in /opt.

o   /sgml Configuration files for SGML and XML

o   /X11 Configuration files for X windows.

In practice most Linux distributions have many more subdirectories of /etc representing optional startup and configuration requirements.

- /home (Optional) Contains user home directories. Each user has a subdirectory under home with the same name as his/her user name. Although FHS calls this optional, in fact it is almost universal among Unix systems. The contents of subdirectories under /home is of course variable.

- /lib Contains those shared library images needed to boot the system and run the commands in the root filesystem—i.e., the binaries in /bin and /sbin. In Linux systems /lib has a subdirectory, /modules, that contains kernel loadable modules.

- /mnt Provides a convenient place to temporarily mount a filesystem.

- /opt Contains optional add-in software packages. Each package has its own subdirectory under /opt.

- /root Home directory for the root user. This is not a requirement of FHS but is normally accepted practice and highly recommended.

- /sbin Contains binaries of utilities essential for system administration such as booting, recovering, restoring or repairing the system. These utilities are only used by the system administrator and normal users should not need /sbin in their path.

- /tmp Temporary files.

- /usr Secondary hierarchy, see below.

■ /var Variable data. Includes spool directories and files, administrative and logging data, and transient and temporary files. Basically, system-wide data that changes during the course of operation. There are a number of subdirectories under /var.

The /usr hierarchy

/usr is a secondary hierarchy that contains user-oriented files. Figure 2.8 shows the subdirectories under /usr. Several of these subdirectories mirror functionality at the root. Perhaps the most interesting subdirectory of /usr is /src for source code. This is where the Linux source is generally installed. You may in fact have sources for several Linux kernels installed in /src under subdirectories with names of the form:

linux-<version number>-ext

You would then have a logical link named linux pointing to the kernel version you're currently working with.

```
/usr          Secondary Hierarchy
  ├── X11R6     X Window system, version 11 release 6 (optional)
  ├── bin       Most user command binaries
  ├── games     Games and educational binaries (optional)
  ├── include   Header files included by C programs
  ├── lib       Libraries
  ├── local     Local hierarchy
  ├── sbin      Non-vital system binaries
  ├── share     Architecture-independent data
  ├── src       Source code (optional)
  │
```

Figure 2-8: /usr Hierarchy

The Shell

One of the last things that happens as a Linux system boots up is to invoke the command interpreter program known as the *shell*. Its primary job is to parse commands you enter at the console and execute the corresponding program. But the shell is much more than just a simple command interpreter. It incorporates a powerful, expressive interpretive programming language of its own. Through a combination of the shell script language and existing utility programs it is quite possible to produce very sophisticated applications without ever writing a line of C code. In fact this is the general philosophy of Unix programming. Start with a set of simple utility programs and link them together through the shell scripting language.

There are in fact several shell programs in common use. They all serve the same basic purpose yet differ in details of syntax and features. The most popular are:

- Bourne Again Shell – bash

- TC Shell – tch

- Z Shell – zsh

The subject of shell programming is worthy of a book in itself and in fact there are seven books on shell programming listed on *Amazon.com*.

Resources

Sobel, Mark G., *A Practical Guide to Linux*. An excellent beginner's guide to Linux and Unix-like systems in general.

tldp.org – The Linux Documentation Project. As the name implies, this is the source for documentation on Linux. You'll find how-to's, in-depth guides, FAQs, man pages, even an on-line magazine, the Linux Gazette. The material is generally well-written and useful.

## CHAPTER 3

The Host Development Environment

In many cases, the target computer on which an embedded application runs is severely limited in terms of its computing resources. It probably doesn't have a full-sized display or keyboard. It may have at most a few megabytes of mass storage in the form of a flash file system, hardly enough to contain a compiler, much less a decent IDE. Thus, embedded development usually requires at least two computers—the target on which the embedded program will run and a development workstation on which the embedded program is written and compiled. Before we begin working with an embedded Linux environment, we'll have to set up an appropriate development workstation.

Any modern PC will work just fine as a development host. Minimum requirements are 32 megabytes of RAM and 2 gigabytes of disk for Linux. Install your favorite Linux distribution, Red Hat, Debian, Mandrake. Modern Linux distributions sport reasonably clean, self-explanatory installation procedures, so I won't duplicate any of that here. A "workstation" class installation is adequate. If it isn't installed by default, you should install the kernel source code.

I do most of my Linux work on a 300-MHz AMD K6 that I call my "lab computer" (see Figure 3-1). It's a dual boot system since I also work in the Microsoft world. I have 2 gigabytes of disk dedicated to Win 95 and 6 gigabytes for Linux, of which about a third is currently used. I run Red Hat version 6.2. Of the several distributions that I've tried, I find 6.2 to be the most stable and the easiest to install and use.

You will need at least one, and preferably two, asynchronous serial ports. You will also need a network interface. OK, you might be able to live without it, but it does make life a lot easier. We'll use a combination of serial and network ports to communicate with the target as well as debug target code.

Figure 3-1: My Linux Development Machine

Cross-Development Tools—the GNU Tool Chain

Not only is the target computer limited in resources, it may be a totally different processor architecture from your (probably) x86-based development workstation. We therefore need a cross-development tool chain that runs on the PC but generates code for a different processor. We also need a tool that will help us debug code running on the target.

GCC

By far the most widely used compiler in the Linux world is GCC, the Gnu Compiler Collection. It was originally called the Gnu C Compiler but the name was changed to reflect its support for more than just the C language. GCC has language front ends for C, C++, Objective C, Fortran, Java and Ada, as well as run-time libraries for these languages.

GCC also supports a wide range of target architectures in addition to the x86. Supported targets include:

- Alpha
- ARM
- M68000
- MIPS
- PowerPC
- SPARC

GCC can run in a number of operating environments including Linux and other variants of Unix. There are even versions that run under DOS and Windows.

GCC can be built to run on one architecture (a PC, for example) while generating code for a different architecture (a PowerPC perhaps). This is the essence of cross development.

GDB

GDB stands for the "Gnu DeBugger." This is a powerful source-level debugging package that lets you see what's going on inside your program. You can step through the code, set breakpoints, examine and change variables and so on. GDB itself is command line driven, making it rather tedious to use. The solution is to front GDB with a graphical front end such as DDD. This combination makes a truly usable source level debugger.

Like GCC, GDB can be built to work with different target architectures.

Configuring and Building the Kernel

One of the neatest things about Linux is that you have the source code. You're free to do whatever you want with it. Most of us have no intention, or need, to dive in and directly hack the kernel sources. But access to the source code does mean that the kernel is highly configurable. That is, you can build a kernel that precisely matches the requirements, or limitations, of your target system.

Now again, if your role is writing applications for Linux, or if you're a Linux system administrator, you may never have to touch the kernel. But as an embedded systems developer, you will most certainly have to build a new kernel, probably several times, either for the workstation or the target environment. Fortunately, the process of configuring and building a kernel is fairly straightforward. My experience has been that building a new kernel is a great confidence-building exercise, especially if you're new to Linux.

When building a new kernel, it's generally a good idea to start with clean, "virgin," source downloaded from www.kernel.org, particularly if you need to patch the kernel to support some added functionality such as RT Linux. Patch files are generally based on clean kernel sources and may not execute correctly on sources that have been otherwise modified.

The remainder of this section details and explains the various steps required to build an executable kernel image.

Where Is the Source Code?

It's not at all unusual to have several versions of the Linux kernel and corresponding source code on your system. How do you cope with these multiple versions? There is a naming strategy that makes this complexity manageable.

Generally, Linux sources are installed as subdirectories of /usr/src. The subdirectories usually get names of the form linux-<version_number>-<additional_features>. <version_number> identifies the base version of the kernel as obtained from kernel.org and looks something like this: 2.4.18. The first number is the "version," in this case 2. This number increments only when truly major architectural changes are made to the kernel and its API. We've been at 2 for something like six years now.

The second number, 4, is called the "patch level" and identifies subversions where the kernel API may differ but the differences aren't enough to justify a full version change. An interesting policy about patch levels is that even numbers represent stable, production kernels and odd numbers identify the kernel currently under development. If you're a hard-core kernel hacker you'll want to keep up with the latest odd-numbered patch level. Those of us who want something that works will stick to the latest even number.

The final number, 18 in this example, is called the "sub-level." Basically it represents bug fixes and enhancements that don't change the kernel API. Applications built to a specific patch level should, in principle, run on any sub-level.

<additional_features> is a way of identifying patches made to a stock kernel to support additional functionality. For example, the book CD includes a kernel called linux-2.4.18-rthal5. This represents sub-level 18 of the stock 2.4 kernel patched to support the RTAI hardware abstraction layer version 5.

Whatever its subdirectory name, the kernel you're currently working with is identified by a symbolic link in /usr/src called linux.

The kernel source tree

Needless to say, the kernel encompasses a very large number of files—C sources, headers, makefiles, scripts, etc. So not surprisingly, there's a standard directory tree to organize these files in a manageable fashion. Figure 3-2 shows the kernel source tree starting at /usr/src/linux. The directories are as follows:

Documentation – Pretty much self-explanatory. This is a collection of text files describing various aspects of the kernel, problems, "gotchas," and so on. There are several subdirectories under Documentation for topics that require more extensive explanations.

arch – All architecture-dependent code is contained in subdirectories of arch. Each architecture has a directory under arch with its own subdirectory structure. The executable kernel image will end up in arch/

<architecture>/boot. An environment variable in the makefile points to the appropriate target architecture directory.

drivers – Device driver code. Under drivers is a set of subdirectories for various devices and classes of device.

fs – Filesystems. Under fs is a set of directories for each type of filesystem that Linux supports.

include – Header files. Among the subdirectories of include are a set of the form "asm-<arch>" where <arch> is the same name as the subdirectory of arch that represents a specific architecture. These directories hold header files containing in-line assembly code, which of course is architecture-dependent. A link named asm points to the subdirectory of the target you're building for.

init – Contains two files; main.c and version.c.

ipc – Code to support Unix System 5 Inter-Process Communication mechanisms such as semaphores, message passing and shared memory.

kernel – This is the heart of the matter. Most of the basic architecture-independent kernel code that doesn't fit in any other category is here.

lib – Several utility functions that are collected into a library.

```
/usr/src/linux
        ├── Documentation    Important information about the kernel
        ├── arch             Subdirectories providing architecture-specific code
        ├── drivers          Subdirectories providing device driver code
        ├── fs               Subdirectories supporting various file sytems
        ├── include          Header files for kernel source code
        ├── init             Code for kernel initialization
        ├── ipc              Code for inter-process communication mechanisms
        ├── kernel           Architecture-independent kernel source code
        ├── lib              Utility functions
        ├── mm               Code for memory management
        ├── net              Code for network support
        └── scripts          Configuration and build support
```

Figure 3-2

mm – Memory management functions.

net – Network support. Subdirectories under net contain code supporting various networking protocols.

scripts – Text files and shell scripts that support the configuration and build process.

Configuring the Kernel—make config, menuconfig, xconfig

usr/src/linux contains a standard make file, Makefile, with a very large number of make targets. The process of building a kernel begins by invoking one of the three make targets that carry out the configuration process. make config starts a text-based script that sequentially steps you through each configuration option. For each option you have either three or four choices. The three choices are: "y" (yes), "n" (no) and "?" (ask for help). The default choice is shown in upper case. Some options have a fourth choice, "m", which means build this feature as a loadable kernel module rather than build it into the kernel image. Figure 3-3 shows an excerpt from the make config dialog.

Most options include help text that is generally "helpful" (see Figure 3-4).

Figure 3-3: make config Dialog

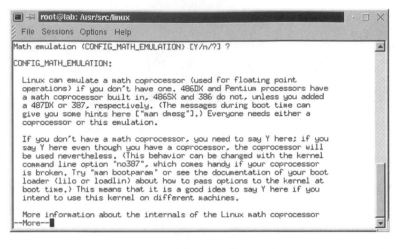

Figure 3-4: make config Help Text

The problem with make config is that it's just downright tedious. Typically you'll only be changing a very few options and leaving the rest in their default state. But make config forces you to step through each and every one. Personally I've never used make config and I wouldn't recommend it.

make menuconfig brings up the pseudo-graphical screen shown in Figure 3-5. Here the configuration options are grouped into categories and you only need to visit the categories of options you need to change. The interface is well-explained and reasonably intuitive. The same help text is available as with

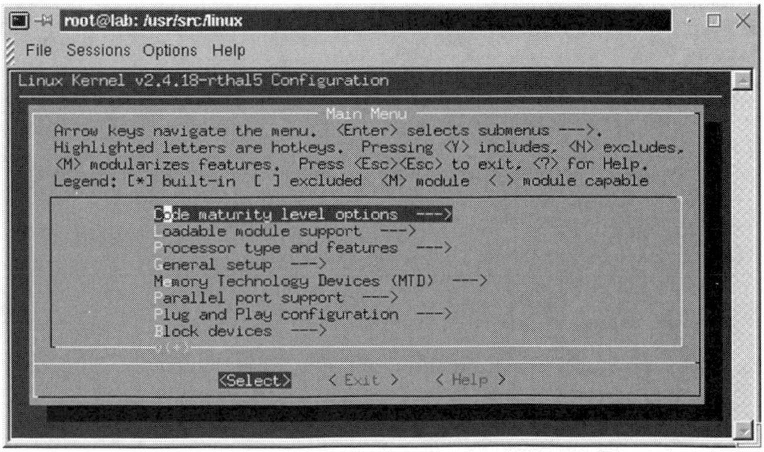

Figure 3-5: make menuconfig Main Menu

make config. When you exit the main menu, you are given the option of saving the new configuration.

But my choice for overall ease of use is make xconfig. This brings up an X Windows-based menu as shown in Figure 3-6. Now you can see all the option categories at once and navigate with the mouse. Of course you must be running X Windows to use this option.

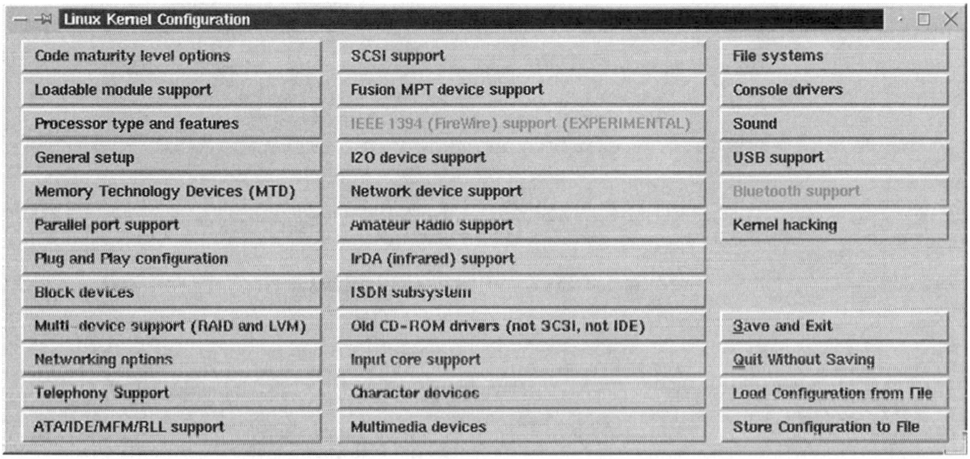

Linux Kernel Configuration		
Code maturity level options	SCSI support	File systems
Loadable module support	Fusion MPT device support	Console drivers
Processor type and features	IEEE 1394 (FireWire) support (EXPERIMENTAL)	Sound
General setup	I2O device support	USB support
Memory Technology Devices (MTD)	Network device support	Bluetooth support
Parallel port support	Amateur Radio support	Kernel hacking
Plug and Play configuration	IrDA (infrared) support	
Block devices	ISDN subsystem	
Multi-device support (RAID and LVM)	Old CD-ROM drivers (not SCSI, not IDE)	Save and Exit
Networking options	Input core support	Quit Without Saving
Telephony Support	Character devices	Load Configuration from File
ATA/IDE/MFM/RLL support	Multimedia devices	Store Configuration to File

Figure 3-6: make xconfig Main Menu

make xconfig gives you additional flexibility in saving configurations. Instead of always saving it to the standard .config file, you can save it to a named file of your choice and later load that file for further modification or to make it the new configuration.

Figure 3-7 is an excerpt from the General Setup menu. Most of the options have the standard "y", "n" and "help" selections. Options that may be built as kernel modules have "m" in the middle column of selections on the left. Options that aren't available, because some other option was not selected, are grayed out. In this case, "PCI bridge optimization" is considered "experimental" and is not available because we did not select "Prompt for development and/or incomplete code/drivers" in the "Code maturity level options" menu. Some options like "PCI access mode" have a set of allowable values other than yes or no and are represented by drop-down dialogs. Some options take numeric values.

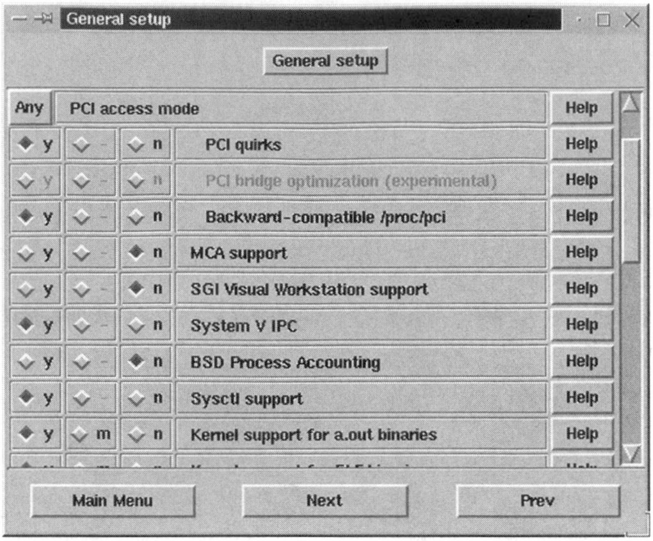

Figure 3-7: xconfig Dialog

Try it out

The best way to become familiar with kernel configuration options is to fire up xconfig and see what's there. So...

    cd /usr/src/linux
    make xconfig

After a bit of program and file building, the menu of Figure 3-6 will appear. Just browse through the submenus and their various sub-submenus to get a feel for the flexibility and richness of features in the Linux kernel. Read the help descriptions.

For the time being leave the configuration menu up while you read the next section.

Behind the Scenes—What's really happening

The information in this section is not essential to the process of building a kernel and you're free to ignore it for now. When you reach the point of developing device drivers or other kernel enhancements (or perhaps hacking the kernel itself), you will need to modify the files that control the configuration process.

All the information in the configuration menus is provided by a set of text files named config.in. These are script files written in Config Language, which looks suspiciously like bash but isn't exactly. The main config.in file is located in linux/arch/$(ARCH) where ARCH is a variable in Makefile identifying the base architecture. Listing 3-1 is an excerpt.

Listing 3-1: Excerpt of config.in

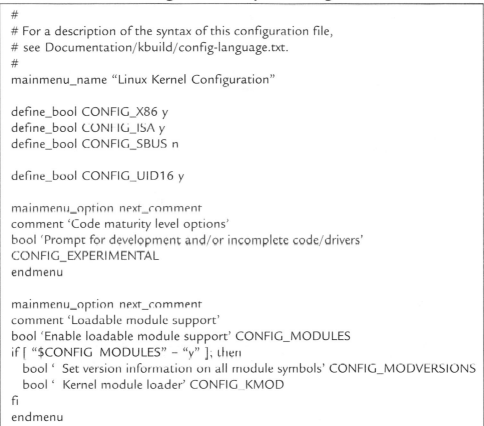

```
#
# For a description of the syntax of this configuration file,
# see Documentation/kbuild/config-language.txt.
#
mainmenu_name "Linux Kernel Configuration"

define_bool CONFIG_X86 y
define_bool CONFIG_ISA y
define_bool CONFIG_SBUS n

define_bool CONFIG_UID16 y

mainmenu_option next_comment
comment 'Code maturity level options'
bool 'Prompt for development and/or incomplete code/drivers'
CONFIG_EXPERIMENTAL
endmenu

mainmenu_option next_comment
comment 'Loadable module support'
bool 'Enable loadable module support' CONFIG_MODULES
if [ "$CONFIG_MODULES" - "y" ]; then
    bool ' Set version information on all module symbols' CONFIG_MODVERSIONS
    bool ' Kernel module loader' CONFIG_KMOD
fi
endmenu
```

Go to linux/arch/i386 and open config.in. Compare the structure of the file with the xconfig menu and the pattern should become fairly clear. Text for each of the menu sections as well as the individual options within those sections is identifiable. There are several types of options:

- bool – The option has two values, "y" or "n".

- tristate – The option has three values, "y", "n" and "m".

- dep_tristate – The option is part of, and dependent on, another option. If the base option has the value "m", then a dependent option is limited to "m" or "n".

- choice – The option has one of the listed values.

- int – The option takes an integer value.

Config Language is actually much more extensive than this simple example would suggest. For more detail, look at linux/Documentation/kbuild/config-language.txt.

Part of each option definition is a symbol of the form CONFIG_*XXXX*. These are the Makefile variables or macros. The whole purpose of configuration is to set these variables. The Makefiles then use these variables to determine which components to include in the kernel and to pass #define symbols to the source code.

Now scroll down toward the end of config.in. You'll eventually come to a section that looks like this:

```
source drivers/mtd/Config.in
source drivers/parport/Config.in
source drivers/pnp/Config.in
source drivers/block/Config.in
source drivers/md/Config.in
if [ "$CONFIG_NET" = "y" ]; then
    source net/Config.in
fi
```

The source keyword is how the configuration menu is extended to incorporate additional, modular features. Note also that configuration options can be conditionally included based on previously selected options.

The help text for configuration options is found in Documentation/Configure.help. The format of this file is quite simple and well documented in the file itself.

The end result of the configuration process, whichever one you choose to use, is a file called .config containing all of the Makefile variables. Listing 3-2 is an excerpt. The options that are not selected remain in the file but are "commented out." The selected options get the appropriate value, "y", "m", a number or an element from a list. .config is included in the Makefile.

Listing 3-2: Excerpt of .config

```
#
# Automatically generated make config: don't edit
#
CONFIG_X86=y
CONFIG_ISA=y
# CONFIG_SBUS is not set
CONFIG_UID16=y


#
# Code maturity level options
#
# CONFIG_EXPERIMENTAL is not set


#
# Loadable module support
#
CONFIG_MODULES=y
# CONFIG_MODVERSIONS is not set
CONFIG_KMOD=y
```

Building the Kernel

Before actually building the kernel, it's important to check the version of gcc, the GNU C compiler, on your system. The command gcc –v will return the version number. The recommended compiler version for 2.4 series kernels is 2.95.3. Later versions (3.x.y) have been known to generate subtle bugs in the kernel even though the build completes successfully. Another version suitable for kernel building is 2.91.66 that was part of the Red Hat 6.2 distribution. This compiler was also known as "kgcc", kernel gcc.

A binary version of gcc 2.95.3 is on the book CD as /usr/local/gcc-2.95.3bin.tar.gz. If needed, install this in /usr/local and add /usr/local/bin to the front of your PATH.

You will need to log in as the root user to actually build the kernel. The build process comprises the following steps:

make dep. This creates two files of dependencies among all of the source and header files for your configuration. The files are .depend and .hdepend.

make clean. Deletes all intermediate files created by a previous build. This insures that *everything* gets built with the current configuration options. You'll find that virtually all Linux makefiles have a clean target.

make bzImage. This is the heart of the matter. This target builds the executable kernel image. Not surprisingly, this takes a while. The resulting compressed kernel image is arch/$(ARCH)/bzImage.

make modules. Builds all of the loadable kernel modules.

make modules_install. Copies the modules to /lib/modules/ <kernel_version> where <kernel_version> is the string identifying the specific kernel you are building.

cp arch/$(ARCH)/bzImage /boot/vmlinuz-<kernel_version>. This copies the kernel image to the /boot directory. As written here, it assumes you're currently in the directory /usr/src/linux.

cp System.map /boot/System.map-<kernel_version>. Copies the system map file to /boot.

Note, incidentally, that the build process is recursive. Every subdirectory in the kernel source tree has its own Makefile dealing with the source files in that directory. The top level Makefile recursively invokes all of the sub Makefiles.

Booting the New Kernel

Most Linux installations incorporate a boot loader, either LILO (Linux Loader) or GRUB (Grand Unified Bootloader) to select the specific kernel or alternate operating system to boot. There's a very good reason for having the ability to boot multiple kernel images. Suppose you build a new kernel and it fails to boot properly. You can always go back and boot a known working image and then try to figure out what went wrong in your new one.

We now need to add our new kernel to the list of kernels recognized by the bootloader. Information about the kernels that LILO can boot is kept in /etc/lilo.conf. Listing 3-3 is an example lilo.conf file. This example shows two different Linux kernel images, one named rtai and the other named linux24, along with an "other" operating system named dos. If no image is specified at the lilo prompt, the default is dos after five seconds.

Listing 3-3: lilo.conf

```
boot = /dev/hda
timeout = 50
linear
prompt
        default = dos
        vga = normal
        read-only
map=/boot/map
install=/boot/boot.b

other = /dev/hda1
        label = dos

image=/boot/vmlinuz-2.2.16-ert
        label=rtai
        read-only
        root=/dev/hda5

image = /boot/vmlinuz-2.4.18-rthal5
        label = linux24
        read-only
        root = /dev/hda5
```

The easiest way to add a new kernel to lilo.conf is to just copy and paste the section for an existing kernel, the four lines beginning with "image =". Then change the image file name and the label to match the kernel you've just built. After saving the file, you must run the command lilo to actually install the boot loader. If you're not logged in as the root user, you may have to type /sbin/lilo because /sbin is typically not in the path of a normal user.

The process with GRUB is much the same. GRUB's boot information is kept in /boot/grub/grub.conf, an example of which is shown in Listing 3-4. In this case, copy and paste the section that begins with "title" up to but not including the next line that begins with "title." Unlike lilo, GRUB does not need to be run to install the boot loader. It gets its information directly from grub.conf at boot time.

Listing 3-4: grub.conf

```
# grub.conf generated by anaconda
#
# Note that you do not have to rerun grub after making changes to this file
# NOTICE:  You have a /boot partition.  This means that
#        all kernel and initrd paths are relative to /boot/, eg.
#        root (hd0,5)
#        kernel /vmlinuz-version ro root=/dev/hda7
#        initrd /initrd-version.img
#boot=/dev/hda
default=1
timeout=10
splashimage=(hd0,5)/grub/splash.xpm.gz
title Red Hat Linux (2.4.7-10)
        root (hd0,5)
        kernel /vmlinuz-2.4.7-10 ro root=/dev/hda7 hdc=ide-scsi
        initrd /initrd-2.4.7-10.img
title DOS
        rootnoverify (hd0,0)
        chainloader +1
```

Summary

This chapter started out discussing the development workstation but quickly digressed into the process of configuring and building a kernel, a process that embedded developers are likely to encounter many times. We'll come back to the workstation in the next chapter when we configure it to work with BlueCat Linux.

Resources

For more details on the process of configuring and building a kernel, look at the files in /usr/src/linux/Documentation/kbuild. Additionally, the following HOW-TOs at *www.tldp.org* may be of interest:

Config-HOWTO – This HOWTO is primarily concerned with how you configure your system once it's built.

Kernel-HOWTO – Provides additional information on the topics covered in this chapter.

## CHAPTER 4

BlueCat Linux

BlueCat Linux is an Open Source Linux distribution developed by LynuxWorks[1]. The BlueCat package provides a basic Linux kernel together with tools and utilities that support cross-development work to put Linux in embedded target devices. The commercial distribution of BlueCat supports a range of target architectures including x86, Power PC, ARM and its derivatives, MIPS and Super H.

LynuxWorks also offers a free downloadable "lite" version of BlueCat that only supports PC-style x86 targets. The book CD includes BlueCat lite in the subdirectory /BlueCat. Because the target is a PC, BlueCat is an ideal tool for learning about, and experimenting with, embedded Linux without investing in a specialized target board.

The "Less Is More" Philosophy

There are a number of companies in addition to LynuxWorks offering Linux distributions that target the embedded marketplace. These include Lineo, RedHat and Monta Vista[2], to name just a few. Many of these embedded toolkits take the approach of encapsulating the Linux development process

---

[1] LynuxWorks didn't actually begin with Linux. They've been around for about 14 years providing a hard real-time version of Unix called LynxOS that is now ABI-compatible (Application Binary Interface) with Linux.

[2] Monta Vista is the only one of these vendors that is not accessible over the Net by the obvious web address. Their website is www.mvista.com.

inside another environment intended to be easier and more intuitive for an embedded developer new to Linux. This is all well and good to the extent that the vendor's environment fully encapsulates the Linux development process. If there are gaps in the encapsulating environment, you may end up having to cope with substantial elements of Linux development as well as the vendor's tools and environment.

LynuxWorks has taken the opposite approach of simply packaging the standard Linux kernel together with a few simple tools to support the target and cross-development environments. The work flow is basically that of standard Linux but the process is well documented in the BlueCat Users' Guide. Minimizing the modifications to the standard Linux code base makes it easier to track changes in Linux.

Installing BlueCat Linux

The lite version takes around 520 megabytes of disk space. Follow these steps to install BlueCat under the /opt directory.

1. Mount the CD, usually under /mnt/cdrom. If you're running a desktop environment like Gnome or KDE (why wouldn't you?), the CD will normally be "automounted" as soon as it's inserted.

2. Go to the /opt directory on your hard disk and make a subdirectory BlueCat.

3. In a shell window, cd /opt/BlueCat.

4. Execute /mnt/cdrom/BlueCat_i386/install.

This completes the BlueCat installation procedure. There is an additional tar file of demo code and documentation in BlueCat_i386/ named bookdemo.tar.gz. Untar this in /opt/BlueCat. It creates subdirectories for several demo projects that we'll be exploring in depth later. There's a minor change to SETUP.sh and some modifications to the osloader demo project. Oddly enough, the downloadable distribution doesn't include the users' guide. So it's also in this file.

The commercial version of BlueCat can also be installed on a Windows host if you're so inclined. This uses the CYGWIN execution environment from Cygnus Solutions (now part of Red Hat).

Exploring BlueCat Linux

Take a look at the directory /opt/BlueCat (Figure 4-1). With three exceptions all of the subdirectories have names identical to those in the standard filesystem hierarchy. They serve the same purpose as the standard directories except they apply to the *target*. They contain executable tools and configuration files that run on the target. These directories become part of the target by including them in the target file system.

```
/opt/BlueCat
            ├── bin        Essential command binaries
            ├── boot       Static files of the boot loader
            ├── cdt        Cross development tools
            ├── demo       Link to target-specific demo directory
            ├── demo.x86   Sample programs for an x86 target
            ├── etc        Target-specific system configuration
            ├── lib        Essential shared libraries and kernel modules
            ├── sbin       Essential system binaries
            ├── usr        Secondary hierarchy
            └── var        Variable data
```

Figure 4-1: BlueCat Linux Directory Structure

The directory cdt/ stands for "Cross Development Tools" and contains the tools necessary to build and debug a target system including, among others, the GNU C compiler and GDB. Under cdt/ you'll find another tree structure that looks suspiciously like the standard filesystem hierarchy. Even though we're building for an x86 target, we have to use the tools in cdt/ rather than the normal set of GNU tools.

The directory link demo/ points to demo.x86/, which in turn contains subdirectories with example systems. In the directory demo/, create a new subdirectory called boot/. Set the access mode for this directory to r-xr-xr-x. The reason for this will become clear later.

The BlueCat Environment

There is a script file in /opt/BlueCat named SETUP.sh. Its job is to set up the appropriate environment for running the BlueCat tools. It exports several environment variables and puts several paths at the beginning of the PATH environment variable so that the BlueCat tool chain is invoked when you make a project. You must run SETUP.sh before working in the BlueCat environment.

However, you must run SETUP.sh using the "dot built-in" command as follows:

> . SETUP.sh

where there is a space between the dot and the file name. Normally, when the bash shell executes a command or script it forks a new process that inherits the exported environment variables of the parent. If we set or change environment variables from the script, the changes only apply to the process executing the script. When it finishes, the changes vanish along with the process.

But bash also has several "built-in" commands, among which is "dot." The dot command runs its script argument in the current process so that the changes take effect in the current process's environment.

> **Try it out**
>
> Try executing SETUP.sh in the normal manner. First, enter the command "set" to see what environment variables are currently defined.
>
> Now run SETUP.sh. The execute permission bit must be set and if your PATH doesn't include the current directory, you'll have to enter ./SETUP.sh. Execute set again. Any changes?
>
> Now use the dot command and enter .SETUP.sh. Run set again.

X86 Target for Blue Cat Linux

BlueCat Linux will run small systems quite nicely even on a 386. This makes it relatively low cost and painless to set up a target for experimenting with BlueCat. Here's your chance to dust off that old 486 box that's been sitting in the closet or serving as a door stop and do something useful with it.

Minimal Requirements

We don't need much to get up and running. In principle we don't even need a keyboard or monitor because Linux can boot and run "headless." Nevertheless it's useful to have a keyboard and display initially to make sure the box is basically running. Here then are the minimal requirements for a BlueCat Linux target:

- 386 or higher motherboard with at least 8 megabytes of RAM.
- Diskette drive, preferably 3.5 inch so it's compatible with your host system. This holds the boot kernel image and root filesystem.
- Serial port.
- Parallel port. We'll use this for experimenting with I/O.
- Network adapter. You can live without it but it really isn't fun. Any common network adapter should suffice. Mine happens to be an NE2000-compatible, BOCALANcard 2000 Plus Combo.

That's it. This configuration is roughly equivalent to a small single-board computer (SBC) that you might use in a real embedded project. The diskette takes the place of a small ROM or Flash memory device. Figure 4-2 shows my ISA and VLB[3] 486 target.

Figure 4-2: My BlueCat target

---

[3] VESA Local Bus. Remember that?

Setup

It's a good idea to have a bootable DOS diskette to do the initial checkout of your target system. If you don't happen to have a bootable diskette, there's an image of one on the book CD named /tools/W95boot.img. Use the DOS utility rawrite in /tools to transfer this image to a diskette. Just start the program and follow the prompts. Note that this process overwrites the entire diskette.

Set up your box with a keyboard and monitor. The first time you power up you should go into BIOS setup, which is usually accomplished by hitting the Del key while the RAM test is in progress. Check that the disk drive configuration matches what you have. I found that POST objected after I removed the hard drive but neglected to change the BIOS setting for Drive C. Exit BIOS setup and boot the DOS diskette. If the boot is successful, you've verified most of the motherboard logic and the diskette drive.

To connect the serial port to your workstation, you will most likely need a "null modem." This is a cable with two 9-pin female D sub connectors wired as shown in Figure 4-3. Transmit Data (TD) and Receive Data (RD) must be swapped so that TD from one end goes to RD on the other end. Likewise the control signals Data Set Ready (DSR) and Data Terminal Ready (DTR) must be swapped. Connect the serial port to COM2 on your workstation.

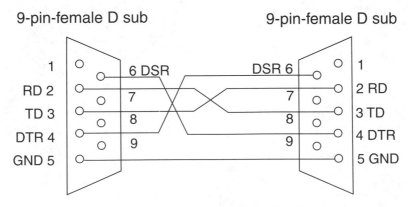

Figure 4-3: Null Modem Cable

W95boot.img contains a simple terminal emulator program, term, to check operation of the serial port. term simply echoes back anything it receives on a serial port at 9600 baud. Usage is:

term <port>

where <port> specifies the serial port number, either 1 or 2. <port> is optional and defaults to 1. Type ^C to exit the program.

You can exercise the target serial port by running the terminal emulator program minicom from your host workstation. But first, you'll need to configure minicom as described in the next section.

If your network connection is twisted pair (10 Base T) then you must use a *crossover* patch cable. Like the null modem, a crossover cable swaps the signal pairs so that the transmitter at one end is connected to the receiver at the other end. There's not much we can do about networking at this point. We'll have to wait until we boot a BlueCat image.

Configuring the Workstation

There are some elements on the host workstation that need to be configured in order to communicate with the target. You need to be the root user in order to make most of the changes discussed here. You can become root user without logging out of your normal user account. In a shell window, enter the command su. Respond to the password prompt with the root password. You're now root user in that shell. To return to your normal user account, enter the command exit.

In KDE you can open a file manager window in root (also called Super User) mode. Open the start menu. Go to System->File manager (Super User Mode). Again you'll be prompted for the root password. The new file manager window has root privileges allowing you to edit and change permissions of files and directories owned by root.

The Terminal Emulator, minicom

minicom is a fairly simple Linux application that emulates a dumb RS-232 terminal through a serial port. The default minicom configuration is through a modem device, /dev/modem. We need to change that to talk directly to one of the PC's serial ports.

In a shell window as root user, enter the command minicom –s. If you're running minicom for the first time you may see the following warning message:

WARNING: Configuration file not found. Using defaults

You will be presented with a configuration menu. Select Serial port setup. Type "A" and replace the word "modem" with "ttyS1". ttyS1 is the device that represents serial port COM2. Type "E" and type "E" again to select 9600 baud. Figure 4-4 shows the final serial port configuration.

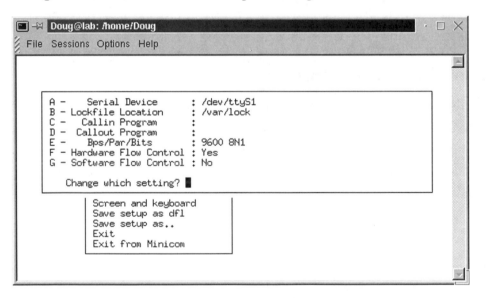

Figure 4-4: minicom Serial Port Settings

Type Enter to exit Serial port setup and then select Modem and dialing. Here we want to delete the modem's Init string and Reset string since they're just in the way on a direct serial connection. Type "A" and backspace through the entire Init string. Type "B" and do the same to the Reset string.

Type Enter to exit Modem and dialing and then select Screen and keyboard. Type "B" once to change the Backspace key from BS to DEL.

Type Enter to exit Screen and keyboard. Select Save setup as… Give the configuration a name, "bluecat" for example. Now select Exit from minicom. The next time you start minicom, enter the command minicom <config> where <config> is the name you just gave to your configuration.

You will probably have to change the permissions on the file /dev/ttyS1 to allow the group and world to read and write the device. Again, you must be root user to do this.

Networking

Your workstation is probably configured to get a network address via DHCP (Dynamic Host Configuration Protocol). But in our network consisting of just the workstation and the target, there is no DHCP server, so we need to specify fixed IP addresses for both ends.

Network configuration parameters are found in /etc/sysconfig/network-scripts/ where you should find a file named something like ifcfg-eth0 that contains the parameters for network adapter 0. Make a copy of this file and name it dhcp-ifcfg-eth0. That way you'll have a DHCP configuration file for future use if needed. Now open the original file with an editor (as root user of course). It should look something like Listing 4-1a. Delete the line BOOTPROTO="dhcp" and add the three new lines shown in Listing 4-1b. The network address 192.168.0 is a good choice here because it's one of a range of network addresses reserved for local networks. For historical reasons, I choose node 11 for my workstation.

```
DEVICE="eth0"
ONBOOT="yes"
BOOTPROTO="dhcp"
```

Listing 4-1a: ifcfg-eth0

```
DEVICE="eth0"
ONBOOT="yes"
IPADDR="192.168.0.11"
NETMASK="255.255.255.0"
BROADCAST="192.168.0.255"
```

Listing 4-1b: revised ifcfg-eth0

We'll use NFS (Network File System) to download images to our target. That means we have to "export" one or more directories on the workstation that the target can mount on its filesystem. Exported directories are specified in the file /etc/exports. Initially this file is present but empty. Open it with an editor and insert the following lines:

```
/opt/BlueCat
/opt/BlueCat/demo/boot 192.168.0.200 (r, no_root_squash)
```

We'll deal with the first line in the next chapter. The second line represents the location from which the target will try to download a kernel image and a root filesystem. The IP address is that of the target. Again, for historical reasons I choose 200 as the node address for the target.

Finally, we have to start the NFS server. This is accomplished with the command /etc/rc.d/init.d/nfs start. You can execute this command from a shell window or, better yet, add it near the end of /etc/rc.d/rc.local. This is the last script executed at boot up.

First Test Program

We're finally ready to download and run something on our target. The best place to start is the osloader demo. This will show how demo systems are organized and built, and also give us a boot loader that can boot other kernel images over the network.

A Demo Directory Roadmap

All of the demo directories are organized with essentially the same files and subdirectories. Take a look at /opt/BlueCat/demo/osloader as an example. In the following descriptions <project> is the name of the project and the subdirectory under demo/. The project directory breaks down as follows:

Directories

local/ – Configuration files specific to this project. These generally end up in /etc or its subdirectories on the target.

src/ – Source files of applications or kernel extensions specific to this project.

Configuration Files

<project>,config – Config file for this project's kernel. When you do make xconfig this file is copied to .config in the kernel source tree. The modifications are made there and then the file is copied back to <project>.config. When you execute make kernel this file is again copied to .config in the kernel source tree.

<project>.spec – This file provides information to build the root file system. The syntax looks like shell commands to create directories and nodes and copy files. It is interpreted by the BlueCat utility mkrootfs.

cl.txt – I added this one. It contains a command line argument for the kernel to redirect the console to ttyS0.

Makefile

The project Makefile supports a number of targets as follows:

xconfig – Copies <project>.config to usr/src/linux/.config and invokes make xconfig from there. Then it copies the resulting .config file back to <project>.config.

kernel – Builds the kernel based on <project>.config.

rootfs – Builds the root file system based on <project>.spec.

kdi – Builds the <project>.kdi target. See below.

this – Builds the custom programs in src/.

all – Builds all of the above targets except xconfig.

clean – Removes all of the target files.

Target Files

These files are the result of building the Makefile targets.

<project>.kernel – Compressed kernel image suitable for booting onto a target board over a network using the BlueCat Linux OS loader. Built by make kernel.

<project>.disk – Compressed kernel image suitable for copying onto a floppy or hard disk. Built by make kernel.

<project>.rfs – Compressed RAM disk root file system image suitable for booting onto a target board over a network using the BlueCat Linux OS loader, or for loading from a floppy disk or hard disk. Built by make rootfs.

<project>.tar – Tar image of the root file system suitable for copying on a hard disk partition or for NFS-mounting. Built by make rootfs.

<project>.kdi – This target contains the compressed kernel image (`.disk`) and the compressed RAM disk root file system (`.rfs`). It is suitable for booting onto a target board from a network using firmware, or programming into ROM/flash memory on the target. Built by make kdi.

Making a Boot Disk

Once the targets are made, it's time to create a bootable diskette. For that we use several invocations of the BlueCat utility mkboot. With a clean, or otherwise unnecessary, diskette in the drive, execute the following set of commands in the directory /opt/BlueCat/demo/osloader:

mkboot –b /dev/fd0 – Writes a boot sector to the diskette.

mkboot –k osloader.disk /dev/fd0 – Copies the kernel image to the diskette.

mkboot –c cl.txt /dev/fd0 – Creates a command line for the kernel using the contents of cl.txt.

mkboot –f osloader.rfs /dev/fd0 – Copies the root filesystem to the diskette.

mkboot –r /dev/fd0 /dev/fd0 – Sets the device node on the target board to mount as the root file system or uncompress the file system image.

After each of these operations mkboot will report on the current status of the diskette. The first time you build a bootable diskette it is important that you do these steps in the order shown. Later on you may find it necessary to modify one or more elements of the root filesystem. You may then execute the mkboot –f command without executing any of the commands that came before it.

However, if you change the kernel and have to rebuild it, you will have to re-execute the mkboot –k command. More than likely the new kernel is larger than the old one and this will overwrite part of the root filesystem. You will then have to re-execute the mkboot –f command.

Executing the Target Image

Take the diskette created in the previous section and insert it into your target system. Turn on power or reboot and while the target is coming up, execute minicom bluecat (or whatever name you saved the minicom configuration under) in a shell window on the workstation. You should see the normal stream of messages that come out of a Linux kernel booting up.

This particular system doesn't do a whole lot. About all it can do, in fact, is to boot another kernel image and root filesystem through NFS using BLOSH, the BlueCat Linux Loader Shell. BLOSH is a shell-like utility with a small set of built-in commands for downloading and executing file images.

When the target finishes booting it presents the BLOSH prompt, ">". Type "set" to see the BLOSH environment variables. These define the target environment parameters such as network interface and IP address, NFS host, kernel image and root filesystem (see Figure 4-5). BLOSH gets these values from a startup file called blosh.rc.

On your workstation go to /opt/BlueCat/demo/bookdemo/shell and copy the files shell.kernel and shell.rfs to ../../boot. Note in Figure 4-5 that these are the files that BLOSH expects to download as the kernel image and root filesystem respectively.

Back in the shell window with minicom running, enter the command boot. This will cause the files shell.kernel and shell.rfs to be downloaded through NFS. The Linux kernel contained in shell.kernel will then start and it will mount the root filesystem contained in shell.rfs.

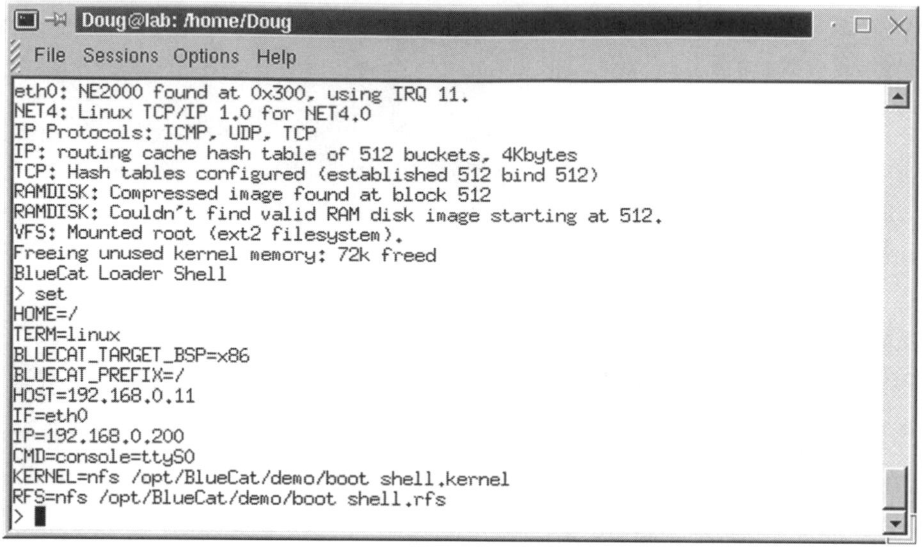

```
┌─────────────────────────────────────────────────────────────┐
│ ■ ─⋈ Doug@lab: /home/Doug                       · □ ✕        │
│ ┊  File  Sessions  Options  Help                             │
│ ┌───────────────────────────────────────────────────────┬─┐ │
│ │eth0: NE2000 found at 0x300, using IRQ 11.             │▲│ │
│ │NET4: Linux TCP/IP 1.0 for NET4.0                      │ │ │
│ │IP Protocols: ICMP, UDP, TCP                           │ │ │
│ │IP: routing cache hash table of 512 buckets, 4Kbytes   │ │ │
│ │TCP: Hash tables configured (established 512 bind 512) │ │ │
│ │RAMDISK: Compressed image found at block 512           │ │ │
│ │RAMDISK: Couldn't find valid RAM disk image starting at 512. │
│ │VFS: Mounted root (ext2 filesystem).                   │ │ │
│ │Freeing unused kernel memory: 72k freed                │ │ │
│ │BlueCat Loader Shell                                   │ │ │
│ │> set                                                  │ │ │
│ │HOME=/                                                 │ │ │
│ │TERM=linux                                             │ │ │
│ │BLUECAT_TARGET_BSP=x86                                 │ │ │
│ │BLUECAT_PREFIX=/                                       │ │ │
│ │HOST=192.168.0.11                                      │ │ │
│ │IF=eth0                                                │ │ │
│ │IP=192.168.0.200                                       │ │ │
│ │CMD=console=ttyS0                                      │ │ │
│ │KERNEL=nfs /opt/BlueCat/demo/boot shell.kernel         │ │ │
│ │RFS=nfs /opt/BlueCat/demo/boot shell.rfs               │ │ │
│ │> ▊                                                    │▼│ │
│ └───────────────────────────────────────────────────────┴─┘ │
└─────────────────────────────────────────────────────────────┘
```

Figure 4-5: BLOSH Environment Variables

When the boot command completes (possibly including some timeout errors that appear to be innocuous), you will be presented with a standard bash prompt. Try out some simple commands like ls. cd to /proc and have a look at some of the /proc files using the cat command.

Finally, take a look at the directory /usr. An ls command executed on this directory will show that it is in fact the same as /opt/BlueCat on the host. Hmmm… How did that happen? Well, if you remember, earlier in this chapter we exported the directory /opt/BlueCat through NFS. On the workstation take a look at /opt/BlueCat/demo/bookdemo/shell/local/ rc.sysint. rc.sysinit is the script file normally interpreted by the init process when the Linux kernel starts.

Near the bottom of rc.sysinit is a pair of mount commands. The first one mounts the /proc filesystem. The second mount command uses NFS to mount the directory /opt /BlueCat on the host as /usr on the target system. So now when we reference a file in /usr on the target, it really references a file on /opt/BlueCat/demo on the host.

Echo the PATH environment variable on the target. Note that it includes /usr/bin and /usr/sbin. These now mapped directly to bin/ and sbin/ in BlueCat/ on the workstation. This means we can execute target programs *stored on the workstation's disk*. We'll exploit this capability to definite advantage in the next chapter.

Resources

Take a look at the BlueCat users' guide, 0443-01-bcl4_users_guide.pdf. It's reasonably good and describes the various utilities that are provided to support the embedded target environment.

For greater insight into what happens as the kernel boots up, check out these HOWTOs at The Linux Documentation Project:

> PowerUp-to-Bash-Prompt-HOWTO
> Linux-Init-HOWTO
> Kernel-HOWTO

## CHAPTER 5

Debugging Embedded Software

The Target Setup

Before diving into the subject of debugging, let's review our setup so far. We have a minimal kernel and minimal root filesystem on diskette that allows us to boot other kernel images and root filesystems over the network via NFS. stdin, stdout and stderr on the target are connected to ttyS0, which in turn is physically connected to ttyS1 on the host. We communicate with the shell running on the target through the terminal emulation program minicom. The boot loader script directs it to boot a kernel image and root filesystem named shell.

The shell kernel is configured to start up a bash shell with a minimal set of command utilities in /bin and /sbin. It also mounts an NFS volume on /usr. The mounted volume is /opt/BlueCat on the host workstation. Consequently, every file in the BlueCat/ directory is accessible from the target. The implication of this, among other things, is that we can execute *on the target* program files that physically reside on the host's filesystem. This allows us to test user space application programs without having to build and download a new root filesystem. And of course, programs running on the target can open files on the NFS-mounted host volume. Again, the console device is redirected to ttyS0 via a command line option to the kernel so that we can communicate with bash running on the target. Figure 5-1 shows graphically how these elements interrelate.

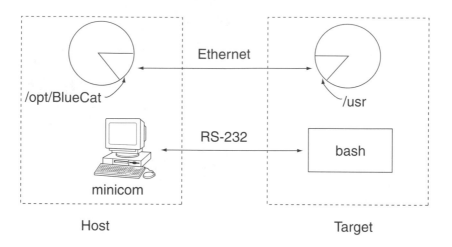

Figure 5-1: The Host and the Target

In principle we could even use the network to communicate with the shell. We simply open a telnet session on the host directed at the target. But BlueCat Lite doesn't include a telnet daemon, so we'll just stick with what we have for now.

The osloader, introduced in the previous chapter, is primarily of value when you're working on kernel code. It saves you the time of writing a diskette every time you build a new kernel image. But once you have a stable kernel, there's no reason not to put that on a diskette to save the additional step of downloading the kernel and filesystem over the net. So at this point it would be useful to put the shell project on a diskette. Refer back to the section *Making a Boot Disk* in the previous chapter for the recipe for making a boot diskette.

GDB

GDB stands for the GNU DeBugger, the source-level debugging package that is part of the GNU toolchain. Like any good debugger, it lets you start your program, insert breakpoints to stop the program at specified points, examine and modify variables and processor registers. Like most Linux utilities, it is command-line driven, making it rather tedious to use. The solution is to front GDB with a graphical front end such as DDD. This combination makes a truly usable source level debugger.

In a typical desktop environment, the target program runs on the same machine as the debugger. But in our embedded environment DDD/GDB runs on the host and the program being debugged runs on the target (see Figure 5-2). Fortunately, GDB implements a serial protocol that allows it to work with remote targets either over an RS-232 link or Ethernet.

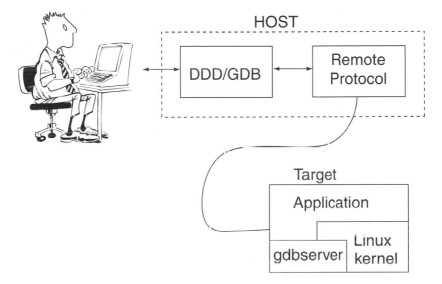

Figure 5-2: Debugging with gdb

There are two approaches to interfacing the target to the GDB serial protocol:

- gdb stubs. A set of functions linked to the target program. gdb stubs is an RS-232-only solution.

- gdbserver. This is a stand-alone program running on the target that, in turn, runs the program to be debugged. The advantage to gdbserver is that it is totally independent of the target program. In other words, the target builds the same regardless of remote debugging. The other major advantage of gdbserver is that it runs over Ethernet.

DDD Front End

DDD provides an X-Windows front end for GDB. You can point-and-click to set breakpoints, examine variables and so on. DDD translates the GUI input into commands for GDB, known in this environment as the "inferior" debugger.

BlueCat lite includes GDB but not DDD or gdbserver. These are on the book CD in /tools. Copy the file gdbserver to /opt/BlueCat/bin. Untar ddd.tar.gz to /usr/local.

Debugging a Sample Program

The easiest way to get started with a software tool is to just try it out. So before getting into the details of remote debugging, let's just try out DDD on the host and play with it a bit. In bookdemo/ is a subdirectory called ddd/. In there is a small program called shellsort.

Build the program with the following command:

```
gcc -g -o shellsort shellsort.c
```

The -g option adds symbol information to the executable for use by GDB.

This program sorts and prints out in ascending order the list of numeric arguments passed to it. For example:

```
./shellsort 4 6 3 1 7 8
1 3 4 6 7 8
```

Most of the time it works. But there's a subtle bug in the program. Try for example:

```
./shellsort 4000 1000 7000 6000 8000
```

The arguments are sorted in the correct order but one of them has been changed. Let's use DDD to find out what's happening. Enter the command:

```
ddd shellsort
```

After a few seconds the display shown in Figure 5-3 appears. The various elements of this display are:

Source Window: Displays the source code around the current execution point.

Command Tool: Buttons for commonly used commands

ToolBar: Contains buttons for commands that require and argument along with a window for entering the argument.

Debugger Console: Lets you enter commands directly to the inferior debugger's command line. There are some operations that just work better at the command line.

Status Line: Shows the current state of DDD and GDB.

We'll encounter other windows as we go along.

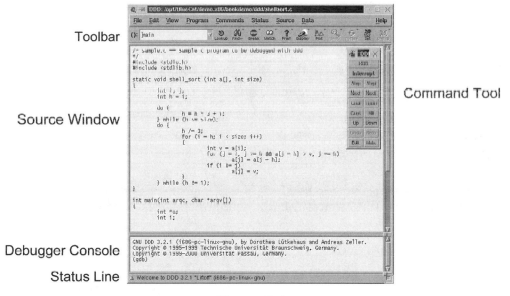

Figure 5-3

The first thing to do now is to set a `Breakpoint` so that the program will stop at a location we are interested in. Click on the blank space to the left of the line that initializes a. The `Argument field` '():' now contains the location ('`sample.c:32`'). Now click on '`Break`' to create a breakpoint at the location in '`()`'. A little red stop sign appears in line 32. Alternatively you can right click on the left side of line 32. A menu appears where one of the options is breakpoint. Selecting this is identical to clicking the breakpoint button in the tool bar.

Now select 'Program->Run' from the file menu. This brings up the Run Program dialog box. In the 'Run With Arguments' window, enter a set of numbers like the ones above that will cause the program to fail. Click on 'Run'. Very shortly the program halts at the breakpoint in line 32. The Debugger Console reports details of the breakpoint and a green arrow appears at line 32 to show the current execution point.

To see the current value of a variable, just place the cursor over it. In about a second a yellow box appears with the current value. Try it with 'argc'. 'a' hasn't been initialized yet so its value is likely to be zero. To execute the current line, click on the 'Next' button on the command tool. The arrow advances to the following line. Now point again to 'a' to see that the value has changed and that 'a' has actually been initialized.

Move the cursor to the left of the call to shell_sort() and right click. Select Set temporary breakpoint and click Cont. The breakpoint we set earlier on line 32 is called a "sticky" breakpoint because it stays there until we specifi- cally remove it. A temporary breakpoint goes away the first time it is hit. Other debuggers call this function "Execute to here."

When the program reaches line 36, the 'a' array has been initialized. To view all values assigned to 'a', enter the following in the argument window:

    a[0]@(argc-1)

and click the Print button. The values of all five elements appear in the Debugger Console. Because 'a' was dynamically allocated, GDB doesn't know how big it is. The "@" notation tells GDB how many elements to list.

Rather than using 'Print' at each stop to see the current value of 'a', you can also display 'a', such that its is automatically displayed. With a[0]@(argc - 1) still showing in the argument window click on 'Display'. The contents of 'a' are now shown in a new window, the Data Window. Click on 'Rotate' to rotate the array horizontally.

Click on Step to step into the shell_sort() function. The execution point arrow moves to the first line of shell_sort() and the Debugger Console shows the arguments passed to it. This is called a "stack frame" display. You can use `Status->Backtrace` to see where you are in the stack as a whole. Selecting a line (or clicking on `Up` and `Down`) will let you move through the stack. Note how the `a` display disappears when its frame is left.

Let's check whether the arguments to shell_sort() are correct. After returning to the lowest frame, enter a[0]@size in the argument field and click on Print:

```
(gdb) print a[0] @ size
$4 = {4000, 1000, 7000, 6000, 8000, 1913}
(gdb)
```

Aha! Where did 1913 come from? We told GDB to print size elements and it printed six. But we only entered five arguments to shellsort. So the size argument being passed to shell_sort() is off by one. To verify this we can change the value of size and see what happens. Select `size` in the source code and click on `Set`. A dialog pops up where you can edit the variable value. Change the value to five and click on Finish to let the program run to completion. Now the result is correct.

Clearly this brief tour has only touched on the highlights. Yet the few commands we've looked at are sufficient for most of your debugging needs.

Setting up for Remote Debugging

Our next step is to move shellsort to the target and debug it there.

gdbserver

gdbserver resides in /opt/BlueCat/bin, which translates to /usr/bin on the target and happens to be in the target's PATH. In the target window, the one running minicom, cd /usr/demo/bookdemo/ddd. Start gdbserver with the command:

```
gdbserver :1234 shellsort 4000 1000 7000 6000 8000
```

The arguments to gdbserver are:

- Network port number preceded by a colon. The normal format is "host:port", where "host" is either a host name or an IP address in the form of a dotted quad, but gdbserver ignores the host portion. The port number is arbitrary as long as it doesn't conflict with another port number in use.

- Target program.

- Arguments to the target program.

gdbserver responds with:

> Process shellsort created; pid = 23

The pid value may be different. This says that gdbserver has loaded the program and is waiting for a debug session to begin on the specified port.

GDB

Start up DDD as above. Now before starting execution, we have to connect GDB to the target. In the Debugger Console window, enter:

> target remote 192.168.0.200:1234

This tells GDB to connect to a remote target, more specifically a network target since the target address is in the form "host":"node". Here the "host" portion is significant as it identifies the target's node address. The "port" number must match the one used when starting gdbserver.

Observe the output from gdbserver showing that the target program has started. Note the corresponding message in the gdb window.

When connected to a remote target, GDB thinks the target program has already started. Set up an initial breakpoint as before, but now use the Cont button instead of Run to proceed. Run through the steps we did above on the workstation to verify that GDB behaves the same.

Note incidentally that gdbserver and shellsort are run from the NFS mounted volume on the host.

The Host as a Debug Environment

Although remote GDB gives us a pretty good window into the behavior of our program on the target, there are still reasons why it might be useful to do initial debugging on our host development machine. To begin with, the host is available as soon as the project starts, probably well before any real target hardware is available or working. The host has a file system that can be used to create test scripts and document test results.

Of course, in order to use this technique, you must have both target and host versions of your operating system. In our current BlueCat environment that's not a problem since both the host and target are x86. But even if the target is different, with Linux it's not a problem since we have the source code by definition.

When you characterize the content of most embedded system software, you will usually find that something like 5% of the code deals directly with the hardware. The rest of the code is independent of the hardware and therefore shouldn't need hardware to test it, provided that you can supply the appropriate stimulus to exercise it.

Unfortunately, all too often the 5% of hardware-dependent code is scattered randomly throughout the entire system. In this case you're virtually forced to use the target for testing because it is so difficult to "abstract out" the hardware dependencies. The key to effective host-level testing is rigorous software structure. Specifically, you need to isolate the hardware-dependent code behind a carefully defined API that deals with the hardware at a higher, more abstract level. This is not hard to do but it does require some planning.

If your hardware-dependent code is carefully isolated behind a well-defined API and confined to one or two source code modules, you can substitute a *simulation* of the hardware-dependent code that uses the host's keyboard and screen for I/O. In effect, the simulated driver "fools" the application into thinking the hardware is really there.

You can now exercise the application by providing stimulus through the keyboard and noting the output on the screen. In later stages of testing you may want to substitute a file-based "script driver" for the screen and keyboard to create reproducible test cases.

Think of this as a "test scaffold" that allows you to exercise the application code in a well-behaved, controlled environment. Among other things, you can simulate limit conditions that might be very difficult to create, and even harder to reproduce, on the target hardware.

The "Thermostat" Example

While the shellsort program might be interesting, it has little to do with real embedded system problems. Here is a sample program closer to the real world that we'll use for the remainder of this chapter and on and off throughout the rest of the book.

cd /opt/BlueCat/demo/bookdemo/thermostat and open the file thermostat.c with an editor. This is a simple implementation of a thermostat. If the measured temperature drops below a specified setpoint, a "heater" turns on. When the temperature rises above the setpoint the heater turns off. In practice, real thermostats incorporate hysteresis that prevents the heater from rapidly switching on and off when the temperature is right at the setpoint. This is implemented in the form of a "deadband" such that the heater turns on when the temperature drops below the setpoint – deadband and doesn't turn off until the temperature reaches setpoint + deadband. Additionally, the program includes an "alarm" that flashes if the temperature exceeds a specified limit.

thermostat.c is fundamentally a state machine that manipulates two digital outputs in response to an analog input and the current state of the program. Note in particular that thermostat.c makes no direct reference to any I/O device. The analog input and digital outputs are virtualized through a set of "device driver" functions. This allows us to have one set of driver functions that work in the host simulation environment (simdrive.c) and another set that work on the target (trgdrive.c).

For our purposes the driver API (Application Programming Interface) is fairly simple and includes the following functions:

- int initAD (void) – Does whatever is necessary to initialize the A/D converter. If the initialization is not successful, it returns a non-zero status value.

- unsigned int readAD (unsigned int channel) – Returns the current value for the channel that was passed in the previous call to readAD() and sets the analog multiplexer to *channel*. This allows for a multi-channel implementation although it's not used here.

- void closeAD (void) – Does whatever is necessary to "clean up" when the program terminates. May not be necessary in some implementations.

- setDigOut (int bitmask) – Turns on the bit(s) specified by bitmask.

- clearDigOut (int bitmask) – Turns off the bit(s) specified by bitmask.

Take a look at simdrive.c. This driver implementation uses a shared memory region to communicate with another process that displays digital outputs on the screen and accepts analog inputs via the keyboard. This functionality is implemented in devices.c. The shared memory region consists of a data structure of type shmem_t (defined in driver.h) that includes fields for an analog input and a set of digital outputs that are assumed connected to LEDs. It also includes a process ID field (pid_t) set to the pid of the devices process that allows the thermostat process to signal when a digital output has changed.

devices creates and initializes the shared memory region. In simdrive, initAD() attaches to the previously created shared memory region. readAD() simply returns the current value of the a2d field in shared memory. The setDigOut() and clearDigOut() functions modify the leds field appropriately and then signal the devices process to update the screen display. Figure 5-4 illustrates the process graphically.

Build both devices and thermostat with the command make sim. To run this example on the host, start up two shell windows. In the first, run devices and in the second run thermostat.s (the .s identifies the simulation version). Change the "A/D in:" input in the devices window and verify that the thermostat responds correctly. Run thermostat.s under DDD to see how the state transitions progress.

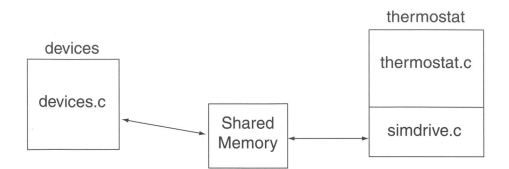

Figure 5-4: Thermostat Simulation

Adding Programmable Setpoint and Limit

As presently designed, the setpoint, limit and deadband for the thermostat are "hardwired" into the program. At the very least, setpoint and limit should be adjustable, preferably remotely. The obvious solution is to use stdin to send simple commands to change these parameters.

If we were building this under DOS, we would probably use the function kbhit() to poll for console input within the main processing loop in thermostat. But Linux is a multitasking system. Polling is tacky! The proper solution is to create an independent thread of execution whose sole job is to wait for console input and then interpret it.

So here's your chance to do some real programming. Modify thermostat.c to incorporate settable setpoint and limit parameters. We'll use a very simple protocol to set parameters consisting of a letter followed by a number. "s" represents the setpoint and "l" the limit. So to change the setpoint you would enter, for example,

s 65<Enter>

This sets the setpoint to 65 degrees.

We'll create a new child process that waits for keyboard input, parses the received string and updates the specified parameter. Refer back to the description of fork() in chapter 2. Add the fork() code to thermostat.c after the call to initAD() and before the while loop. The child process is shown in pseudocode form in Listing 5-1.

Remember though that the child process inherits a *copy* of the parent's data space, not the same data space. So even if the child resides in the same source file as the parent, it sees a different copy of setpoint and limit. Changes made to these variables by the child are not seen by the parent.

The solution is to put the parameters into a data structure and allocate a shared memory region for that structure just as simdrive does for the simulated peripherals. The parent will create the shared memory region and both the parent and child must get and attach to it.

Listing 5-1: Command Parsing Process

```
get and attach shared memory region (see simdrive.c for details)

while (1)
{
        char string[20];
        char *token;

        gets (string);
        token = strtok (string, whitespace);
        value = atoi (strtok (NULL, whitespace));

        switch (token[0]);
        {
                case 's':
                        // update setpoint
                case 'l':
                        // update limit
                default:
                        // ignore
        }
}
```

If you're not familiar with any of the functions shown here, refer to the corresponding man page. If you're not familiar with man pages, see the Resources section below.

Once you have the program built, try it out under DDD. Note that you'll be using the same console device to enter the command string that thermostat uses to write out the current temperature. This results in a certain amount of interference.

Once you're satisfied with the program's operation in the host, the next step is to build it for the target. We'll do that in the next chapter.

There is another, actually somewhat simpler, approach to creating a second thread of execution using Posix Threads. Whereas a process carries with it a complete execution context and full set of protected resources, a thread is a thread of execution only. The only resources a thread owns are code and a stack. The distinction is illustrated graphically in Figure 5-5. Note that multiple threads within a process share the process's data space.

UNIX Process Model

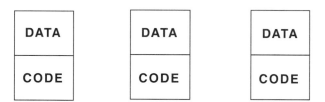

Multi-threaded Process

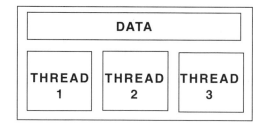

Figure 5-5: "Processes" vs. "Threads"

The thread API is actually more like traditional multitasking programming than the Linux fork() model. That is, you create a thread out of a function by calling a thread creation service. Creating a thread is generally a lower overhead operation than creating a process and threads are often characterized as "light-weight multitasking".

Posix threads is covered in Chapter 11.

Resources

DDD includes an extensive, well-written users' manual in both HTML and PDF. GDB is well covered in the GNU "info" pages.

In addition to GDB and DDD, this chapter introduced several Linux programming concepts. An excellent guide to Unix/Linux programming is

Matthew, Niel and Richard Stone, *Beginning Linux Programming 2nd Edition*, Wrox Press, 1999

Man and Info Pages

The "official" Unix/Linux documentation consists of a large collection of "man" (manual) pages. There is one "page" for each topic of interest. Of course, for complex topics that page may stretch to several hundred lines. To make it easier to find a specific topic the man pages are organized into sections as follows:

Section 1:	User Commands entered at the shell prompt.
Section 2	The kernel API.
Section 3:	C library functions.
Section 4:	Devices. Information on specific peripheral devices.
Section 5:	File formats. Describes the syntax and semantics for the files in /etc.
Section 6:	Games.
Section 7:	Miscellaneous.
Section 8:	System Administration. Shell commands primarily used by the system administrator.

Man pages can be accessed from the command line with:

man <topic>

Of course it's much easier to access the man pages through the KDE desktop. Just click on the help icon in the tool bar. The main help window has a link to the man pages.

The GNU info pages provide more extensive documentation on various elements of the GNU toolchain and other software available from the Free Software Foundation. Like the man pages, info pages can be accessed through the command:

info <topic>

But again, the KDE help window has a link to the info pages.

## CHAPTER 6

Kernel Modules and
Device Drivers

For me, the fun part of embedded programming is seeing the computer interact with its physical environment—i.e., actually do something. In this chapter we'll use the parallel port on our target machine to provide some limited I/O capability for the thermostat program developed in the previous chapter. But first some background.

Kernel Modules

Installable kernel modules offer a very useful way to extend the functionality of the basic Linux kernel and add new features without having to rebuild the kernel. A key feature is that modules may be dynamically loaded when their functionality is required and subsequently unloaded when no longer needed. Modules are particularly useful for things like device drivers and /proc files.

Given the range and diversity of hardware that Linux supports, it would be impractical to build a kernel image that included all of the possible device drivers. Instead the kernel includes only drivers for boot devices and other common hardware such as serial and parallel ports, network devices and so on. Other devices are supported as loadable modules and only the modules needed in a given system are actually loaded.

Loadable modules are unlikely to be used in a production embedded environment because we know in advance exactly what hardware the system must support and so we simply build that support into the kernel image. Nevertheless, modules are still useful when testing a new driver. You don't need to build a new kernel image every time you find and fix a problem in your driver code. Just load it as a module and try it.

Keep in mind that modules execute in Kernel Space at Privilege Level 0 and thus are capable of bringing down the entire system.

A Module Example

cd /opt/BlueCat/demo/bookdemo/driver and open the file hello.c. This is a trivial example of a loadable kernel module. It contains two functions; init_module() and cleanup_module(). Every module must include these two functions. init_module() is called by insmod, the shell command that installs a module. cleanup_module() is called by rmmod, the command that removes a module.

In this example both functions simply print a message on the console using printk, the kernel equivalent of printf. C library functions like printf are intended to run from user space making use of operating system features like redirection. These facilities aren't available to kernel code. Rather than writing directly to the console, printk writes to a circular buffer and then wakes up the klogd process to deal with the message by either printing it to the console and/or writing it to the system log.

Note the "<1>" at the beginning of the printk format strings. This is a *loglevel* that determines whether or not the message appears on the console. Loglevels range from 0 to 7 with lower numbers having higher priority. If the loglevel is numerically less than the kernel integer variable console_loglevel then the message appears on the console. In any case, it also shows up in the file /var/log/messages.

Oddly enough, printk messages do not appear in shell windows running under X windows regardless of the loglevel. You can see what printk did with the command:

> tail /var/log/messages

This prints out the last few lines of the messages log file.

This example also shows how to specify module parameters with the MOD_PARM() macro.

Take a look at the Makefile for hello. Note in particular the two symbols defined by the environment variable K_CFLAGS. The __KERNEL__ symbol is required to compile code that runs as part of the kernel. The MODULE symbol is required for code that will be loaded as a kernel module.

Note also the –c flag in the compiler command line. This means compile only, don't link with any libraries. So the resulting object file is not directly executable and may contain references to external symbols such as printk. How do these external references get resolved? insmod resolves them against the kernel's symbol table, which is loaded in memory as part of the kernel boot process. Furthermore, any nonstatic symbols defined in the module are added to the kernel symbol table and are available for use by subsequently loaded modules. So the only external symbols a module can reference are those built into the kernel image or previously loaded modules. The kernel symbol table is available in /proc/ksyms.

Make hello and try it out. Enter the command

```
insmod ./hello my_string="name" my_int=47
```

If you're running from the command line, i.e. not in X windows, you should see the message printed by init_module(). If you're running a shell from within X windows, in KDE for example, execute tail /var/log/messages. Note, by the way, that insmod doesn't by default look in the current directory. Now try the command lsmod. This lists the currently loaded modules. It also gives a "usage count" for each module and shows what modules depend on other modules. The same information is available in /proc/modules. Enter the command ksyms. This lists the symbols that have been added to the kernel symbol table by the currently loaded modules. Now execute the command rmmod hello. You should see the message printed by cleanup_module(). Finally execute lsmod again and the list should be empty.

"Tainting" the Kernel

If you're running Modutils version 2.4.18 from the book CD, you saw another message when you installed hello:

> Warning: loading ./hello will taint the kernel: no license

> See http://www.tux.org/lkml#export-tainted for information about tainted modules

What the heck does that mean? Apparently, kernel developers were getting tired of trying to cope with bug reports involving kernel modules for which no source was available, that is, modules not released under an Open Source license such as GPL. Their solution to the problem was to invent a MODULE_LICENSE() macro whereby you can declare that a module is indeed Open Source. The format is:

> MODULE_LICENSE ("<approved string>")

Where <approved_string> is one of the ASCII text strings found in linux/include/linux/module.h. Among these, not surprisingly, is "GPL". If you distribute your module in accordance with the terms of an Open Source license such as GPL, then you are permitted to include the corresponding MODULE_LICENSE() invocation in your code and loading your module will not produce any complaints.

If you install a module that produces the above warning and the system subsequently crashes, the crash documentation (core dump) will reveal the fact that a non-Open Source module was loaded. Your kernel has been "tainted" by code that no one has access to. If you submit the crash documentation to the kernel developer group it will be ignored[1].

Add the following line to hello.c just below the two MODULE_PARM() statements:

---

[1] What's not clear to me is how the tainted kernel provision is enforced. A device vendor could very easily include a MODULE_LICENSE() entry in his driver code but still not release the source. What happens when that module causes a kernel fault? I suspect the Open Source community is relying on public approbation to "out" vendors who don't play by the rules. What else is there?

MODULE_LICENSE("GPL");

Remake hello and verify that installing the new version does not generate the warning.

The Role of a Module

As an extension of the kernel, a module's role is to provide some service or set of services to applications. Unlike an application program, a module does not execute on its own. Rather, it patiently waits until its service is invoked by some application program.

But how does that application program gain access to the module's services? That's the role of init_module(). By calling one or more kernel functions, init_module() "registers" the module's "capabilities" with the kernel. In effect the module is saying "Here I am and here's what I can do."

Figure 6-1 illustrates graphically how this registration process works. init_module() calls some register_capability() function passing as an argument a pointer to a structure containing pointers to functions within the module. The register_capability() function puts the pointer argument into

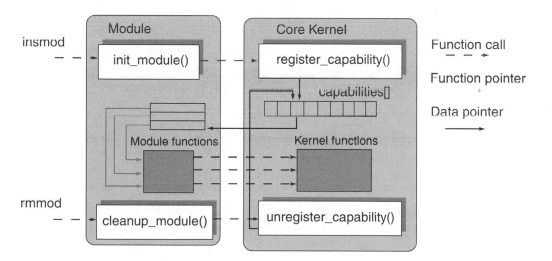

Figure 6-1: Linking a Module to the Kernel[2]

---

[2] Rubini, Allesandro and Jonathan Corbet, *Linux Device Drivers, 2nd Ed*, p. 18

an internal "capabilities" data structure. The system then defines a protocol for how applications get access to the information in the capabilities data structure through system calls.

A specific example of a capabilities structure is chrdevs, the table of character device drivers. The specific kernel function that manages chrdevs is register_chrdev(). The way in which an application program gains access to the services of a character device driver is through the open() system call. This is described in more detail below.

If a module is no longer needed and is removed with the rmmod command, cleanup_module() should "unregister" the module's capabilities, freeing up the entry in the capabilities data structure.

What's a Device Driver Anyway?

There are probably as many definitions of a *device driver* as there are programmers who write them. Fundamentally, a device driver is simply a way to "abstract out" the details of peripheral hardware so the application programmer doesn't have to worry about them.

In simple systems a driver may be nothing more than a library of functions linked directly to the application. In general-purpose operating systems, device drivers are often independently loaded programs that communicate with applications through some OS-specific protocol. In multitasking systems like Linux the driver should be capable of establishing multiple "channels" to the device originating from different application processes. In all cases though the driver is described in terms of an API that defines the services the driver is expected to support.

The device driver paradigm takes on additional significance in a protected mode environment such as Linux. There are two reasons for this. First, User Space application code is normally not allowed to execute I/O instructions. This can only be done in Kernel Space at Privilege Level 0. So a set of device driver functions linked directly to the application simply won't work. The driver must execute in Kernel Space. Actually, there are some hoops you can jump through to allow I/O access from User Space but it's better to avoid them.

The second problem is that User Space is swappable. This means that while an I/O operation is in process, the user's data buffer could get swapped out to disk. And when that page gets swapped back in, it will very likely be at a different physical address. So data to be transferred to or from a peripheral device must reside in Kernel Space, which is not swappable. The driver then has to take care of transferring that data between Kernel Space and User Space.

Linux Device Drivers

Unix, and by extension Linux, divides the world of peripheral devices into three categories:

- Character
- Block
- Network

The principal distinction between character and block is that the latter, such as disks, are randomly accessible—that is, you can move back and forth within a stream of characters. Furthermore, data on block devices is usually transferred in one or more blocks at a time and prefetched and cached. With character devices the stream moves in one direction only. You can't, for example, go back and reread a character from a serial port. Block devices generally have a filesystem associated with them, whereas character devices don't.

Network devices are different in that they handle "packets" of data for multiple protocol clients rather than a "stream" of data for a single client. Furthermore, data arrives at a network device asynchronously from sources outside the system. These differences necessitate a different interface between the kernel and the device driver.

The /dev Directory

While other OSes treat devices as files, Linux goes one step further in actually creating a directory for devices. Typically this is /dev. In a shell window on the workstation, cd /dev and ls –l hd*. You'll get a very large number of entries. Notice that the first character in the flags field for all these entries is "b". This indicates that the directory entry represents a "block" device. In fact all of these entries are hard disk devices.

Between the Group field and the date stamp, where the file size normally appears, is a pair of numbers separated by a comma. These are referred to as the "major device number" and the "minor device number" respectively. The major device number turns out to be an index into a table of device driver pointers, either chrdev[] for character devices or blkdev[] for block devices. So, in effect, the major device number identifies the device driver. The minor device number is used only by the driver itself to distinguish among possibly different types of devices the driver can handle. The major and minor device numbers are each eight bits.

Now do ls –l tty*. The first character in the flags field is now "c" indicating that these are character devices. Character devices and block devices each have their own table of device driver pointers. Links allow the creation of logical devices that can then be mapped to system-specific physical devices. Try ls –l mouse. In my system this is a link to psaux, the device for a PS2-style mouse.

Entries in /dev are created with the mknod command as, for example,

```
mknod  /dev/ttyS1  c  4  65
```

This creates a /dev entry named ttyS1. It's a character device with major device number 4 and minor device number 65. There's no reference to a device driver here. All we've done with mknod is create an entry in the filesystem so application programs can "open" the device. But before that can happen we'll have to "register" a character device driver with major number 4 to fill the appropriate entry in the chrdev[] table.

The Low Level I/O API

The set of User Space system functions closest to the device drivers is termed "low level I/O." For the kind of device we're dealing with in this chapter, low level I/O is probably the most useful. For general file transfers the "stream" I/O package makes more sense.

Figure 6-2 shows the basic elements of the low level I/O API.

- OPEN. Establishes a connection between the calling process and the file or device. path is the directory entry to be opened. oflags is a bitwise set of flags specifying access mode and must include one of the following:

O_RDONLY	Open for read-only.
O_WRONLY	Open for write-only.
O_RDWR	Open for both reading and writing.

  Additionally oflags may include one or more of the following modes:

O_APPEND	Place written data at the end of the file.
O_TRUNC	Set the file length to zero, discarding existing .contents.
O_CREAT	Create the file if necessary. Requires the function call with three arguments where mode is the initial permissions.

  If OPEN is successful it returns a non-negative integer representing a "file descriptor." This value is then used in all subsequent I/O operations to identify this specific connection.

- READ and WRITE. These functions transfer data between the process and the file or device. filedes is the file descriptor returned by OPEN. buf is a pointer to the data to be transferred and count is the size of buf in bytes. If the return value is non-negative, it is the number of bytes actually transferred, which may be less than count.

- CLOSE. When a process is finished with a particular device or file, it can close the connection, which invalidates the file descriptor and frees up any resources to be used for another process/file connection. It is good practice to close any unneeded connections because there is typically a limited number of file descriptors available.

- IOCTL. This is the "escape hatch" to deal with specific device idiosyncrasies. For example, a serial port has a baud rate and may have a modem attached to it. The manner in which these features are controlled is specific to the device. So each device driver can establish its own protocol for the IOCTL function.

```
int open (const char *path, int oflags);
int open (const char *path, int oflags, mode_t mode);
size_t read (int filedes, void *buf, size_t count);
size_t write (int filedes, void *buf, size_t count);
int close (int filedes);
int ioctl (int filedes, int cmd, ...);
```

Figure 6-2: Stream I/O API

A characteristic of most Linux system calls is that, in case of an error, the function return value is −1 and doesn't directly indicate the source of the error. The actual error code is placed in the global variable errno. So you should always test the function return for a negative value and then inspect errno to find out what really happened. Or, better yet, call perror( ), which prints a sensible error message on the console.

There are a few other low level I/O functions but they're not particularly relevant to this discussion.

Internal Driver Structure

The easiest way to get a feel for the structure of a device driver is to take a look at the code for a relatively simple one. In opt/BlueCat/demo/ bookdemo/thermostat is a simple driver called parport.c. This simply reads and writes the PC's parallel port in a way that supports our thermostat example. Open the file with an editor.

Basically, a driver consists of a set of functions that mirror the functions in the low level API. However these functions are called by the kernel in Kernel Space in response to an application program calling a low level I/O function.

init_module() and cleanup_module()

Let's start near the end of the file with the init_module() function. The first thing the driver must do is gain exclusive access to the relevant I/O ports. First we check that the range of interest is not being used with a call to check_region(). If that succeeds, we get the ports with a call to request_region().

For the parallel port BASE represents the data register and is normally 0x378. BASE + 1 is the status register. For the thermostat example, the data port will drive a pair of LEDs representing the heater and the alarm. The status port will monitor a set of pushbuttons that allow us to raise or lower the measured "temperature."

Finally, we register the device driver with a call to register_chrdev(). The arguments are:

- A major device number, in this case 6. 6 happens to be the major device number normally used by the parallel port device driver.

- A pointer to a file_operations structure. This contains pointers to the functions that implement the driver API for this device. The file_operations structure is just above init_module(). Our driver is quite simple and only implements four of the driver functions. The kernel takes some sensible default action for all of the driver functions that are specified as NULL. register_chrdev() puts the file_operations pointer in chrdev[major], in this case chrdev[6].

- A name. The principal purpose of the name is to create a mnemonic association with the major device number. The name shows up in the /proc/devices file.

If init_module() succeeds, the return value is zero. A non-zero return value indicates an error condition and the module is not loaded. Note that if init_module() fails, any resources successfully allocated up to the point of failure must be returned before exiting. In this case, failure to register the device requires us to release the I/O region.

As might be expected, cleanup_module() simply reverses the actions of init_module(). It unregisters the device and releases the I/O region.

open() and release()

Move back to the top of the file around line 28. The kernel calls parport_open() when an application calls open() specifying /dev/parport as the path. The arguments to open both represent a "file" but in a slightly different way. struct inode represents a file on a disk or a physical device. Among the fields in inode are the major and minor device numbers. struct file represents an *open* file. Every open file in the system has an associated file structure that is created by the kernel on open() and passed to every function that operates on the file until the final close(). struct file maintains information like the access mode, readable or writable, and the current file position. There's even a void *private_data field that allows the driver to add its own information if necessary.

Neither inode nor file is relevant to our simple parallel port driver. All parport_open() has to do is increment the module use count with the macro MOD_INC_USE_COUNT. The purpose of the module use count is to prevent a module from being removed if it is being used by an application or by another module. Thus, rmmod will only remove a module if its use count is zero.[3]

Interestingly enough, the driver equivalent of close is called release. In this case, all we have to do is decrement the module use count.

read() and write()

Moving down the file we come to the interesting functions, parport_read() and parport_write(). The arguments to read and write are:

- A struct file (see above).

---

[3] I have found this to be a minor annoyance in debugging kernel modules. Sometimes the bug is that the use count doesn't get properly incremented or decremented and ends up non-zero with no way to zero it out. The only solution is to reboot. I think rmmod ought to have an override to allow removing a module with a non-zero use count.

- A pointer to a data buffer *in User Space*.

- A count representing the size of the buffer.

- A pointer to a file position argument. If the function is operating on a file, then any data transferred will change the file position. It is the function's responsibility to update this field appropriately.

parport_read() is a little unusual in that it will never return more than two bytes of data regardless of the count value. We invert the status byte because a pressed button reads as a logic 0 on the status lines but we would like to see a button press appear as a 1 in the program. However the MSB of the status register is already inverted, so we don't invert it here.

The port registers are read into a local char array in Kernel Space. copy_to_user() copies this local array to the application's buffer in User Space.

parport_write() takes the more conventional approach of writing out an arbitrary size buffer even though only the last character written will be visible on the LEDs. If count is greater than 2, we dynamically allocate a Kernel Space buffer of count bytes. Normally count is 1 and it's rather silly to allocate a buffer of one byte so we just use a local variable. In either case, copy_from_user() copies the application's buffer into the Kernel Space buffer. Then just write it out and return the dynamically allocated buffer if necessary.

Figure 6-3 attempts to capture graphically the essence of driver processing for a write operation. The user process calls write(). This invokes the kernel through an INT instruction where the write operation is conveyed as an index into the syscall[] table. The filedes argument contains, among other things, a major device number so that the kernel's write function can access the driver's file_operations structure by indexing into chrdev. The kernel calls the driver's write function, which copies the User Space buffer into Kernel Space.

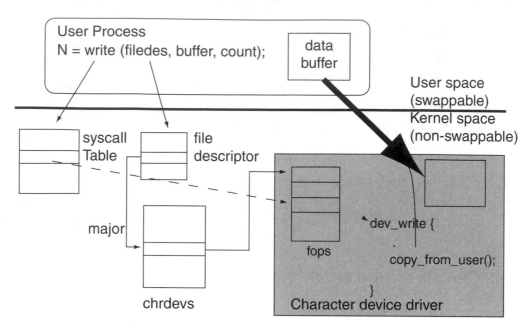

Figure 6-3: Driver Processing — Write

Building and Running the Driver

Build the driver with make parport. Now before we can actually load and exercise the driver, we have to create a device node on the target filesystem. cd ../shell and edit shell.spec. Below the line that begins "mknod /dev/ttyS1" add the line

```
mknod /dev/parportc 6 0
```

Now rebuild the root filesystem with make rootfs and copy it to your diskette with mkboot –f shell.rfs /dev/fd0. Reboot your target.

In the target window on your workstation, ls –l /dev. The parport device should show up. Now cat /proc/ioports. Several I/O port ranges will be listed but you should *not* see one that begins "0378". cat /proc/devices. You'll see a number of character devices but you shouldn't see one for major number 6.

cd /usr/demo/bookdemo/thermostat and insmod ./parport.o. If the module installed successfully, repeat the cat commands for /proc/ioports and /proc/devices. parport should show up in both listings. The driver is now installed and ready to use. But before we do that, we need to take a little side trip into the exciting world of hardware.

The Hardware

OK, it's time to dust off the old soldering iron. Yes, I know, this is a book about software and operating systems, but any successful embedded developer needs to get his or her hands dirty once in a while.

Figure 6-4 is a schematic of the device we need to build. It's pretty simple really, consisting of just two LEDs and three pushbuttons connected to a DB-25P connector, the mate to the parallel port connector on the target PC. The LEDs are connected to the least significant bits of the data port, the pushbuttons to the most significant bits of the status port. Feel free to add more LEDs to display the rest of the data port on pins 4 through 9. There are also two additional status bits available on pins 13 and 15.

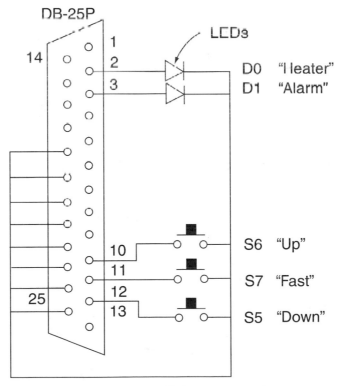

Figure 6-4: Parallel Port Interface

The schematic assumes the use of LEDs with built-in dropping resistors so they can be safely connected to a 5-volt source. If you use LEDs without the built-in dropping resistor, you should add a resistor of between 1k and 4.7k in series with each LED.

As shown in Figure 6-4, the three pushbuttons represent "Up," "Down" and "Fast." In the context of thermostat, pressing "Up" causes the measured temperature to increase by two degrees each time it is read. Pressing "Down" decreases the temperature by two degrees. Pressing "Fast" together with either "Up" or "Down" changes the temperature by five degrees per scan.

This is admittedly somewhat awkward, but is the result of not having enough digital input bits. If we had, say, six bits we could attach a DIP switch to each bit and just read a number directly.

The Target Version of Thermostat

Now we're ready to try building and running thermostat on the target with "real" hardware. Take a look at trgdrive.c. It has the same set of functions as simdrive.c but these interact with the parallel port through the parport device driver. This is a set of "wrapper" functions that translate the raw I/O of the parallel port into something that makes sense to the application. Later on, if we were to plug a real A/D converter into the target, we would simply write a new set of trgdrive functions that talk to its device driver with absolutely no change to thermostat.c.

make target. With the parport module loaded, execute ./thermostat.t (.t for the target version) in the target's shell window. Now play with the pushbuttons and watch the heater and alarm respond accordingly.

Debugging Kernel Code

Kernel Space code presents some problems in terms of debugging. To begin with, DDD and GDB rely on kernel services. If we stop the kernel, at a breakpoint for example, those services aren't available. Consequently some other approaches tend to be more useful for kernel and module code.

printk

Personally I've never been a big fan of the printf debugging strategy. On occasion it's proven useful but more often than not the print statement just isn't in the right place. So you have to move it, rebuild the code and try again. I find it much easier to probe around the code and data with a high quality source level debugger.

Nevertheless, it seems that many Unix/Linux programmers rely heavily on printf and its kernel equivalent printk. At the kernel level this makes good sense. Keep in mind, of course, that printk exacts a performance hit.

While printk statements are useful during development, they should probably be taken out before shipping production code. Of course as soon as you take them out, someone will discover a bug or you'll add a new feature and you'll need them back again. An easy way to manage this problem is to encapsulate the printk's in a macro as illustrated in Listing 6-1.

Listing 6-1

```
#ifdef DEBUG
#define PDEBUG(fmt, args...) printk (<1> fmt, ## args)
#else
#define PDEBUG(fmt, args...)     // nothing
#endif
```

While you're debugging, define the DEBUG macro on the compiler command line. When you build production code, DEBUG is not defined and the printk statements are compiled out.

/proc Files

The /proc filesystem serves as a window into the kernel and its device drivers. In fact many Linux utilities, lsmod for example, get their information from /proc files. Some device drivers export internal information via /proc files and so can yours.

/proc files are harder to implement than printk statements but the advantage is that you only get the information when you ask for it. Once a /proc file is created, there's virtually no run time overhead until you actually ask for the data. printk statements on the other hand always execute.

A simple read-only /proc file can be implemented with the two function calls shown in Listing 6-2. A module that uses /proc must include <linux/proc_fs.h>. The function create_proc_read_entry() creates a new entry in the /proc directory. This would typically be called from init_module(). The arguments are:

- Name. The name of the new file.

- File permissions. Who's allowed to read it. The value 0 is treated as a special case that allows everyone to read the file.

- struct proc_dir_entry. Where the file shows up in the /proc hierarchy. A NULL value puts the file in /proc.

- Read function. The function that will be called to read the file.

- Private data. An argument that will be passed to the read function.

Listing 6-2

```
#include <linux/proc_fs.h>

struct proc_dir_entry *create_proc_read_entry (char *name,
        mode_t mode, struct proc_dir_entry *base,
        read_proc_t *read_proc, void *data);

int read_proc (char *page, char **start, off_t offset,
        int count, int *eof, void *data);
```

The read_proc() function is called as a result of some process invoking read() on the /proc file. Its arguments are:

- page. Pointer to a page (4k bytes) of memory allocated by the kernel. read_proc() writes its data here.

- start. Where in page read_proc() starts writing data. If read_proc() returns less than a page of data you can ignore this and just start writing at the beginning.

- offset. This many bytes of page were written by a previous call to read_proc(). If read_proc() returns less than a page of data you can ignore this.

- count. The size of page.

- eof. If you're writing more than one page, set this non-zero when you're finished. Otherwise you can ignore it.

- data. The private data element passed to create_proc_read_entry().

Like any good read function, read_proc() returns the number of bytes read.

As a trivial example, we might want to see how many times the read and write functions in parport are called. Listing 6-3 shows how this might be done with a /proc file.

kgdb

Actually it is possible to debug the kernel with GDB. The basic requirement is that you have two systems; a host machine on which GDB runs and a target machine on which the kernel code you're debugging runs. These two machines are connected through a serial port. Well, that's exactly the setup we've got.

kgdb is available from kgdb.sourceforge.net. It's a kernel patch that adds the following features to the kernel:

- Gdb stub. This is the heart of the debugger. It services requests coming from GDB on the host. It has control when the kernel is stopped by the debugger.

- Modifications to fault handlers. The kernel gives control to the debugger when an unexpected fault occurs.

- Serial communication. Provides an interface to gdb stub using a serial driver in the kernel.

Listing 6-3

```
Add to parport.c

#include <linux/proc_fs.h>

int read_count, write_count;    // These are incremented each time the
                                // read and write functions are called.

int parport_read_proc (char *page, char **start, off_t offset, int count, int
*eof, void *data)
{
  return sprintf (page, "parport. Read calls: %d, Write calls: %d\n",
      read_count, write_count);
}

In init_module()

  create_proc_read_entry ("parport", 0, NULL, parport_read_proc, NULL);
```

Download the appropriate patch file and apply it to your kernel source. Next, configure the kernel. Enable "Remote (serial) kernel debugging with gdb" in the Kernel hacking section. This will enable a few more configuration options that modify kgdb's behavior. Before building you'll need to add "-g" to the compiler flags in the kernel Rules.make file. This adds debugging information to the kernel image for gdb.

You can also use kgdb for debugging modules. There's a shell script on sourceforge that jumps through all the hoops to load a module for debugging.

Building Your Driver into the Kernel

In a production embedded environment there is little reason at all to use loadable kernel modules. So when you have your new driver working you'll probably want to build it into the kernel executable image. This means integrating your driver into the kernel's source tree. It's not as hard as you might think.

To begin with, you need to move your driver source files into the kernel source tree. Assuming your source code really is a "device driver," it probably belongs somewhere under usr/src/linux/drivers. In the absence of any definitive documentation or recommendations on this issue, I recommend you create your own directory under usr/src/linux/drivers. I called mine usr/src/linux/drivers/doug. Copy your driver source code to this directory and create a Makefile like the one shown in Listing 6-4. This is for a driver that consists of a single object file. If your driver contains multiple object files, look at the Makefile in drivers/bluetooth/ as an example.

Listing 6-4

```
#
# Makefile for Doug's parport driver.
#
parport.o:  parport.c

include  $(TOPDIR)/Rules.make
```

The source file requires a couple minor modifications that cause init_module() to be called as part of the kernel boot process. Add the include file linux/init.h. In the declaration of init_module() add "__init" just before init_module so that it looks like this:

int __init init_module (void)

__init causes the function to be compiled into a special segment of initialization functions that are called as part of the boot up process. Once boot up is complete, this segment can be discarded and the memory reused.

At the end of the file add the following line:

__initcall(init_module);

This causes a pointer to init_module() to be added to a table of initialization functions that are called at boot up. The upshot is that you don't have to modify main.c to add a new initialization function every time you add a new driver. It happens automatically.

Also, we no longer need cleanup_module() because the device will never be removed.

Next you'll need to make your device directory visible to the Makefile in drivers/. Open usr/src/linux/drivers/Makefile with an editor and take a look at the section starting at line 15. There are a number of lines of the form:

subdir-$(CONFIG_xxxx) += yyyy

where CONFIG_xxxx represents an environment variable set by the make xconfig process and yyyy is a subdirectory of drivers/. All of these environ-ment variables end up with one of three values:

"y" Yes. Build this feature into the kernel.

"n" No. Don't build this feature into the kernel.

"m" Module. Build this feature as a kernel loadable module.

So the Makefile is building a list of all the subdirectories under drivers/ that need to be built. Move down to line 58 after a section bracketed by "#ifdef __bluecat__". If you want to add your driver directory unconditionally, add a line like this:

subdir-y += <your directory>

You can also make your driver a kernel configuration option by defining an environment variable of the form CONFIG_yyyy. Then, instead of the line in the previous paragraph, you would add to the Makefile a line of the form:

subdir-$(CONFIG_yyyy) += <name>

But now you need to add an entry into the configuration menus. Refer back to the section in Chapter 3, *Behind the Scenes—What's really happening*.

Finally, you'll need to make your device driver object file visible to the kernel's Makefile. Open usr/src/linux/Makefile with an editor and scroll down to a section around line 175 that starts with "DRIVERS-n :=". Follow-ing this is a large set of lines of the form:

DRIVER-$(CONFIG_xxxx) += drivers/<subdir>/yyyy.o

where again CONFIG_xxxx represents an environment variable set by the make xconfig process and yyyy.o is an object file in some subdirectory of drivers.

When this section of the Makefile completes, the environment variable DRIVER-y contains a list of all the driver object files that should be linked into the kernel image. DRIVER-m is a list of the driver object files that should be built as kernel loadable modules and DRIVER-n represents those drivers that aren't needed at all. It's not at all clear to me why the kernel build process needs to know about drivers that won't be built.

Move down to the end of this section of the Makefile around line 254, where it says "DRIVERS := $(DRIVERS-y)", and add one of the following lines depending on whether or not your driver is conditionally included:

DRIVER-y += /drivers/<your_subdir>/<driver>.o

DRIVER-$(CONFIG_yyyy) += /drivers/<your_subdir >/<driver>.o

Try it out. Follow the instructions above and then execute make kernel in bookdemo/shell/. Boot the new kernel either from a diskette or with osloader. You should see parport show up in both /proc/ioports and /proc/devices.

In a true production kernel you would also want to remove loadable module support from the kernel to reduce size.

An Alternative—uCLinux

Much of the complexity and runtime overhead of conventional Linux device drivers is the result of the protected memory environment. In calling a device driver, the system switches from User Space to Kernel Space and back again. Data to be transferred to/from a peripheral device must be copied from/to User Space. All of this chews up run time.

uCLinux is a variant of Linux designed to run on processors without memory management. The uC stands for microcontroller. Without memory management there's no User Space and Kernel Space, no privilege levels. Everything runs in one flat memory space effectively at Privilege Level 0. There's no virtual memory and no swapping.

The lack of privilege levels and memory protection has several implications. First of all, any process can directly execute I/O instructions. This further implies that device drivers aren't really necessary. Of course as a structuring and abstraction tool drivers are still useful, but now they can be as simple as functions linked directly to a process executable image.

The other major consequence is that any process can bring down the whole system, just like kernel loadable modules in conventional Linux.

uClinux has been ported to a number of Motorola microcontrollers including the Dragonball (M68EZ328) and other 68k derivatives as well as Coldfire. Other ports include the Intel i960, ARM7TDMI and NEC V850E. For more information on uClinux, go to www.uclinux.org.

Arcturus Networks offers a number of hardware development kits based on uClinux. Check them out at www.arcturusnetworks.com.

Resources

The subject of module and device driver programming is way more extensive than we've been able to cover here. Hopefully, this introduction has piqued your interest and you'll want to pursue the topic further. An excellent book on the topic is:

Rubini, Alessandro and Jonathan Corbet, *Linux Device Drivers, 2ⁿᵈ Ed.*, O'Reilly, 2001.

In fact, I would go so far as to say this is one of the best computer science books I've read. It's very readable and quite thorough.

CHAPTER 7

Embedded Networking

It's a "net-centric" world, as the marketers like to say, so it's time to turn our attention to network programming in the embedded space. Linux, as a Unix derivative, has extensive support for networking.

Sockets

The "socket" interface, first introduced in the Berkeley versions of Unix, forms the basis for most network programming in Unix systems. Sockets are a generalization of the Unix file access mechanism that provides an endpoint for communication either across a network or within a single computer. A socket can also be thought of as extension of the named pipe concept that explicitly supports a client/server model wherein multiple clients may be attached to a single server.

The principal difference between file descriptors and sockets is that a file descriptor is bound to a specific file or device when the application calls open(), whereas sockets can be created without binding them to a specific destination. The application can choose to supply a destination address each time it uses the socket, for example when sending datagrams, or it can bind the destination to the socket to avoid repeatedly specifying the destination, for example when using TCP.

Both the client and server may exist on the same machine. This simplifies the process of building client/server applications. You can test both ends on the same machine before distributing the application across a network. By convention, network address 127.0.0.1 is a "local loopback" device. Processes can use this address in exactly the same way they use other network addresses.

> **Try it out**
>
> Execute the command /sbin/ifconfig. This will list the properties and current status of network devices in your system. You should see at least two entries: one for eth0, the Ethernet interface and the other for lo, the local loopback device.
>
> This command should work on both your development host and your target with similar results. ifconfig, with appropriate arguments, is also the command that sets network interface properties.

The Server Process

Figure 7-1 illustrates the basic steps that the server process goes through to establish communication. We start by creating a socket and then bind() it to a name or destination address. For local sockets, the name is a file system entry often in /tmp or /usr/tmp. For network sockets it is a *service identifier* consisting of a "dotted quad" Internet address (as in 192.168.0.11, for example) and a protocol port number. Clients use this name to access the service.

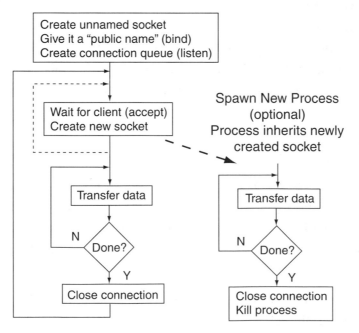

Figure 7-1: The Server Process

Next, the server creates a connection queue with the listen() service and then waits for client connection requests with the accept() service. When a connection request is received successfully, accept() returns a new socket, which is then used for this connection's data transfer. The server now transfers data using standard read() and write() calls that use the socket descriptor in the same manner as a file descriptor. When the transaction is complete, the newly created socket is closed.

The server may very well spawn a new process to service the connection while it goes back and waits for additional client requests. This allows a server to serve multiple clients simultaneously. Each client request spawns a new process with its own socket.

The Client Process

Figure 7-2 shows the client side of the transaction. The client begins by creating a socket and naming it to match the server's publicly advertised name. Next, it attempts to connect() to the server. If the connection request succeeds, the client proceeds to transfer data using read() and write() calls with the socket descriptor. When the transaction is complete, the client closes the socket.

If the server spawned a new process to serve this client, that process should go away when the client closes the connection.

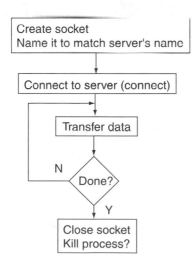

Figure 7-2: The Client Process

Socket Attributes

The socket system call creates a socket and returns a descriptor for later use in accessing the socket.

```
#include <sys/socket.h>
int socket (int domain, int type, int protocol);
```

A socket is characterized by three attributes that determine how communication takes place. The *domain* specifies the communication medium. The most commonly used domains are PF_UNIX for local file system sockets and PF_INET for Internet connections. The "PF" here stands for Protocol Family.

The domain determines the format of the socket name or address. For PF_INET, the address is AF_INET and is in the form of a dotted quad. Here "AF" stands for Address Family. Generally there is a 1 to 1 correspondence between AF_ values and PF_ values. A network computer may support many different network services. A specific service is identified by a "port number". Established network services like ftp, http, etc have defined port numbers, usually below 1024. Local services may use port numbers above 1023.

Some domains, PF_INET for example, offer alternate communication mechanisms. SOCK_STREAM is a sequenced, reliable, connection-based, two-way byte stream. This is the basis for TCP and is the default for PF_INET domain sockets. SOCK_DGRAM is a *datagram* service. It is used to send relatively small messages with no guarantee that they will be delivered or that they won't be reordered by the network. This is the basis of UDP. SOCK_RAW allows a process to access the IP protocol directly. This can be useful for implementing new protocols directly in User Space.

The protocol is usually determined by the socket domain and you don't have a choice. So the protocol argument is usually zero.

A Simple Example

The Server

cd /opt/BlueCat/demo/bookdemo/network and open the file netserve.c. First we create a server_socket that uses streams. Next we need to bind this socket to a specific network address. That requires filling in a sockaddr_in structure, server_addr. The function inet_aton() takes a string containing a network address as its first argument, converts it to a binary number and stores it in the location specified by the second argument, in this case the appropriate field of server_addr. Oddly enough, inet_aton() returns *zero* if it succeeds. In this example, the network address is passed in through the compile-time symbol SERVER so that we can build the server to run either locally through the loopback device or across the network.

The port number is 16 bits and is arbitrarily set to 4242. The function htons() is one of a small family of functions that solves the problem of transferring binary data between computer architectures with different byte-ordering policies. The Internet has established a standard "network byte order," which happens to be Big Endian. All binary data is expected to be in network byte order when it reaches the network. htons() translates a short (16-bit) integer from "host byte order," whatever that happens to be, to network byte order. There is a companion function, ntohs() that translates back from network byte order to host order. Then there is a corresponding pair of functions that do the same translations on long (32-bit) integers.[1]

Now we *bind* server_socket to server_addr with the bind() function. Finally, we create a queue for incoming connection requests with the listen() function. A queue length of one should be sufficient in this case because there's only one client that will be connecting to this server.

Now we're ready to *accept* connection requests. The arguments to accept() are:

- The socket descriptor.

- A pointer to a sockaddr structure that accept() will fill in.

- A pointer to an integer that currently holds the length of the structure in argument 2. accept() will modify this if the length of the client's address structure is shorter.

---

[1] Try to guess the names of the long functions.

accept( ) blocks until a connection request arrives. The return value is a socket descriptor to be used for data transfers to/from this client. In this example the server simply echoes back text strings received from the client until the incoming string begins with "q".

The Client

Now look at netclient.c. netclient determines at run time whether it is connecting to a server locally or across the network. We start by creating a socket and an address structure in the same manner as in the server. Then we *connect* to the server by calling connect( ). The arguments are:

- The socket descriptor.

- A pointer to the sockaddr structure containing the address of the server we want to connect to.

- The length of the sockaddr structure.

When connect( ) returns we're ready to transfer data. The client prompts for a text string, writes this string to the socket and waits to read a response. The process terminates when the first character of the input string is "q".

> **Try it out**
>
> To build the client and server to run on the local host, do:
>
>   make client
>   make server
>
> Open another terminal window. In this window cd to the network/ directory and execute ./netserve. Go back to the original window and execute ./netclient. Type in a few strings and watch what happens. To terminate both processes, enter a string that begins with "q" ("quit" for example).

Both the server and client are built with debugging information on, so you can run either or both of them under DDD.

Next we'll want to run netserve on the target with netclient running on the host. Execute the command:

    make server SERVER=REMOTE

In the terminal window connected to the target (the one running minicom), cd to the network/ directory and execute ./netserve. Back in the original window execute ./netclient remote.

A Remote Thermostat

Moving on to a more practical example, our thermostat may very well end up in a distributed industrial environment where the current temperature must be reported to a remote monitoring station and setpoint and limit need to be remotely settable. Naturally we'll want to do that over a network. The network/ directory includes a "network-ready" version of thermostat. And if you had difficulty with the programming assignment in Chapter 5 to make the parameters settable through the serial port, this version of thermostat follows the same basic strategy.

Open network/thermostat.c. This file ends up creating, at a minimum, three processes:

- The basic thermostat process that controls the heater and alarm.
- A net server that accepts connections on behalf of clients that need to know the current temperature or change operational parameters.
- A monitor process created for each network connection accepted by the server. This process parses and acts on commands sent by the client.

Start down in main() at line 195 where we get and attach a shared memory region for the thermostat's operational parameters and data. After initializing the operational parameters this process forks. The parent process goes on to run the thermostat as originally defined back in Chapter 5. The child process invokes a function called server().

Moving up to around line 115, server() is pretty much a duplication of the code we saw earlier in netserve.c. It creates and names a socket, binds to the socket and sets up a connection queue. Then it waits to accept connections on behalf of network clients. The difference is that when server() accepts a network connection, it forks a new monitor() process to service that connection. It then goes back to wait for additional network connections.

This means that the thermostat is capable of responding to multiple clients from anywhere in the network. In a real-world application, you would probably impose additional restrictions such that any client could request the current temperature, but only "trusted" clients would be allowed to change the operational parameters such as setpoint and limit. This could be implemented, for example, by creating two network servers at different port numbers. One server would simply supply current temperature values. The other, responding only to trusted clients, would allow for changing the operational parameters. The implementation of this functionality is left as the proverbial "exercise for the student."

The monitor() process begins at line 57. The first thing it does is get and attach the shared memory region created by the main thermostat process. Then it just reads and acts on commands coming in on the client_socket. Note the use of the client_socket integer variable. It is a global variable in the file thermostat.c. monitor() gets the value of client_socket at the time that the server() process forks. Remember that the value of client_socket seen by monitor() is a *copy* and so from the viewpoint of monitor() it doesn't change, no matter what subsequently happens in server().

The server() process goes back and waits for another client connection. When a new client request arrives, accept() will return a different value to be stored in client_socket. This is the value seen by the new monitor() process when server() forks.

monitor() adds some commands on top of what we discussed in Chapter 5:

- "d" <number> sets the deadband.

- "?"sends back the current temperature.

- "q" terminates the connection and kills the corresponding monitor process.

We might have structured monitor() to automatically send the current temperature to the client at regular intervals in the same way the thermostat currently reports temperature to stdout. But that would have necessitated a different client strategy. The strategy of asking for the current temperature allows us to use netclient as presently constituted to access the thermostat. Nevertheless, feel free to play around with an implementation of both monitor( ) and netclient that automatically returns the current temperature at client-specified intervals.

The Makefile in network/ includes both sim and target targets to build the networked thermostat in both the simulation and target environments. To build the simulation version simply execute

make sim

To run the simulation version you'll need three terminal windows as follows:

1. Runs devices from the thermostat/ directory.

2. Runs thermostat.s from the network/ directory.

3. Runs netclient from the network/ directory.

thermostat is built with the –g compiler flag so you can use DDD to investigate its operation.

When you're happy with the simulation version, you can build the target version of thermostat with

make target SERVER=REMOTE

Before running thermostat.t on the target, make sure the parport device driver is loaded.

Embedded Web Servers

Another approach to network programming is to embedded a web server in your device. This makes it accessible, in principle, from a web browser anywhere on the Internet. This is not the place to delve into web programming but it's worth taking a look at, because BlueCat includes a demo called showcase that incorporates an Apache web server.

The BlueCat documentation says this demo should run in 8 MB of RAM. But I couldn't get it to run on my 8 MB target and various attempts to lower the memory requirement, such as loading library modules from the NFS mount, ultimately proved unsuccessful. So I temporarily "borrowed" a 64 MB box to run this demo.

Once you have a target with sufficient memory—16 MB should suffice—boot the osloader diskette we made back in Chapter 4. Remember that BLOSH (the BlueCat Loader Shell) expects to boot a kernel image and root filesystem from /opt/BlueCat/demo/boot on the workstation. So you'll need to copy showcase.kernel and showcase.rfs from the showcase/ directory under demo/ to the boot/ directory.

Also remember that BLOSH expects to load a kernel image named shell.kernel and a root filesystem named shell.rfs. You have three choices here:

1. You can rename the files in boot/ from showcase.* to shell.*, over-writing what was previously there.

2. You can change the BLOSH environment variables on the target after osloader has booted.

3. You can edit the blosh.rc file in osloader/local/ and rebuild the root filesystem for osloader.

Showcase requires one more magic incantation. It requires more than the default ram disk space. This is handled with another command line argument to the kernel:

```
ramdisk_size=8192
```

This is in addition to the existing "console=" argument. So the full CMD environment variable should read:

```
console=ttyS0 ramdisk_size=8192
```

If showcase boots successfully, the last console output you see before the bash prompt is:

```
Starting Apache server ...
```

This indicates that an http server has been started in the background and is waiting to accept http connection requests.

On your host, fire up the Netscape browser and open:

192.168.0.200/index.html

You should see a marketing pitch for BlueCat Linux.

Behind the Scenes

There's a lot going on here so let's take a look. The showcase target filesystem includes almost no user space utilities like ls or cat, so we'll have to infer a lot of what's happening by looking at the local/ directory and reviewing the file showcase.spec that specifies the root filesystem.

Under showcase/local are three subdirectories. etc/ contains the usual complement of configuration files, conf/ contains configuration files for apache, and html.x86/ contains a set of files for the web page you displayed earlier.

Open showcase.spec. The first thing to notice is that it is heavily dependent on the environment variable BLUECAT_TARGET_BSP. In fact, before you rebuild the root filesystem, you will need to set BLUECAT_TARGET_BSP=x86 and export it. Just about everything related to the web server is concentrated in /etc/httpd. This is where the conf/ directory goes. /etc/httpd contains a link called modules that points to a library of modules for apache and a link called logs that points to a directory under /var/logs. The html files go in /home/httpd/html. The file conf/access.conf has a pointer to this directory.

Take a look at local/etc/.bashrc.x86. This becomes just .bashrc on the target and is the default script executed when bash starts up. It's actually pretty straightforward. All it does is configure the network interface and then start httpd in the background.

Apache is rather extensive and may be overkill in a small embedded application. But it's a start. There are many small web servers scattered around the Internet. Just do a Google search on the phrase "small web server."

Resources

Linux Network Administrators' Guide, available from the Linux Documentation Project, www.tldp.org. Not just for administrators, this is a quite complete, and quite readable tutorial on a wide range of networking issues.

Comer, Douglas, *Internetworking with TCP/IP, Vols. 1 and 2*, Prentice-Hall. This is the classic reference on TCP/IP, currently in its fourth edition. Highly recommended if you want to understand the inner workings of the Internet.

## CHAPTER 8

Introduction to Real-time Programming

Fundamentally, "real-time" means that a program must respond to *events* in its environment within a specified deadline. Such systems are said to be *event-driven* and can be characterized in terms of *latency*, where latency is defined as the time interval from when an event occurs until the time the system takes action in response to that event. In a general-purpose operating system such as Windows or Unix, latency is of little concern. As a user of Windows, you probably couldn't tell the difference if the system responds to a key press in 20 milliseconds or 220 milliseconds. And if the system happens to be doing something else, it may take two seconds to respond. You may or may not notice, but it happens rarely enough that you probably won't be too upset.

Real-time, on the other hand, demands an upper limit on latency, also called the *scheduling deadline*. Real-time systems can be roughly divided into two major classes: *hard* real-time and *soft* real-time. The distinction is that in hard real-time, the system absolutely must meet its scheduling deadline each and every time. Failure to meet the deadline may have catastrophic consequences including loss of life. A fly-by-wire aircraft control system is an example of hard real-time. The control algorithms depend on regular sampling intervals. If sampling is delayed, the algorithm could become unstable.

Consider the aforementioned fly-by-wire system. Suppose the system senses that the plane is losing altitude. It responds by increasing power to the engines, which will reduce the rate of descent. The control algorithm is programmed with a specific relationship between power level and rate of climb or descent for that specific aircraft. Now suppose that the next sample is delayed, maybe because Windows is busy putting up the paper clip icon. The next sample will report a higher altitude than what would have been reported if the sample had been taken at the correct time. So the algorithm erroneously reduces engine power to compensate. At the next sample time the reported altitude will be lower, but may be too low if again the sample time was delayed.

In soft real-time, the scheduling deadline is more of a goal than an absolute requirement. We expect the system to meet its deadline most of the time, but nothing particularly bad happens if it's occasionally late. Failure to meet the deadline simply results in degraded performance without catastrophic consequences. The automated teller network is a good example of soft real-time. Is the ATM network real-time? You bet it is! When you put your ATM card into the machine, you expect a response within a couple of seconds. But if it should take longer, the worst that happens is you get impatient.

Many systems exhibit both kinds of behavior. That is, some parts are hard and some parts are soft.

Polling vs. Interrupts

Real-time systems are said to be "event-driven," meaning that a primary function of the system is to respond to "events" that occur in the system's environment. How does the program respond to events? There are two fundamental approaches. The first is *polling* as illustrated in Listing 8-1. The program begins with some initialization and then enters an infinite loop that tests each possible event to which the system must respond. For each event that is set, the program invokes the appropriate servicing function.

Listing 8-1: The Polling Loop

```
int main (void)
{
sys_init();
while (TRUE)
{
        if (event_1)
            service_event_1();
        if (event_2)
            service_event_2();

            .

            .

        if (event_n)
            service_event_n();
    }
}
```

This strategy is simple to implement and quite adequate for small systems with relatively loose response time requirements. But there are some obvious problems.

- The response time to an event varies widely depending on where in the loop the program is when the event occurs. For example, if event_1 occurs immediately before the if (event_1) statement is executed, the response time is very short. However if it occurs immediately *after* the test, the program must go through the entire loop before servicing event_1.

- As a corollary, response time is also a function of how many events happen to be set at the same time and consequently get serviced in the same pass through the loop.

- All events are treated as having equal priority.

- As new features, hence new events, are added to the system, the loop gets longer and so does the response time.

The second approach, making use of *interrupts,* is much more efficient and, perhaps not surprisingly, more difficult to program. The idea of the interrupt is that the occurrence of an event "interrupts" the current flow of instruction execution and invokes another stream of instructions that services the event as illustrated in Figure 8-1. When servicing is complete, control returns to where the original instruction stream was interrupted. Servicing the event happens "right now" and doesn't have to wait for the main program to "get around to it." The instruction stream that services the event is called an "Interrupt Service Routine" or ISR.

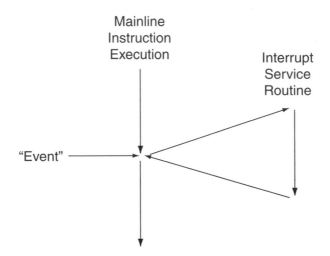

Figure 8-1: The Interrupt

Most modern processors implement three distinct types of interrupt:

- The INT instruction, sometimes called a TRAP. This is like a subroutine call with one important difference as we'll see shortly.

- Processor exceptions. Fault conditions like divide by 0 or illegal memory reference can be handled through the interrupt mechanism.

These two forms of interrupt are *synchronous* with respect to instruction execution. That is, INT *is* an instruction and fault exceptions are the direct result of instruction execution. The third type of interrupt is generated by events that occur external to the processor. These are generated by the input/output hardware and occur *asynchronously* with respect to instruction execution.

It is the asynchronous external interrupt that:

- Maximizes the performance and throughput of computing systems.

- Creates the most problems and frustrations for programmers.

Many years ago someone wrote in *Computer Magazine* of the IEEE Computer Society that "The invention of the interrupt was perhaps the greatest disaster in the history of computer science."

Most processors utilize a similar interrupt scheme. Figure 8-2 shows how the Intel x86 architecture does it. The first 1k bytes of memory[1] are reserved for an *Interrupt Vector Table*. Each vector is four bytes representing the FAR address (segment:offset) of an Interrupt Service Routine. Some of these vectors have specific meanings defined by the processor architecture. For example, vector 0 is the divide-by-0 exception, vector 3 is a breakpoint (a single-byte INT instruction), vector 13 is the infamous General Protection Fault, and so on.

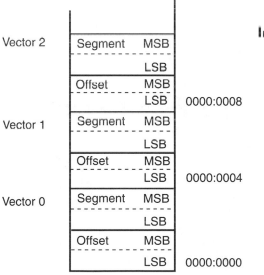

**Figure 8-2:
Interrupts — The Vector Table**

---

[1] This explanation is based on "real mode" memory just to keep it simple.

Some vectors are reserved for external hardware interrupts. In the PC, vectors 8 to 15 and 0x70 to 0x77 are reserved for hardware.

All other vectors are accessible via 2-byte INT instructions, where the second byte is the vector (or interrupt) number. System software usually establishes conventions concerning many of these vectors. For example the PC BIOS uses several interrupts for hardware services and Linux uses INT 0x80 to invoke kernel services.

Figure 8-3 illustrates the basic process of interrupt execution using the Linux INT 0x80 as an example:

- The processor saves the current Program Counter (PC) and Code Segment (CS) on the stack along with the Processor Status Word (PSW).

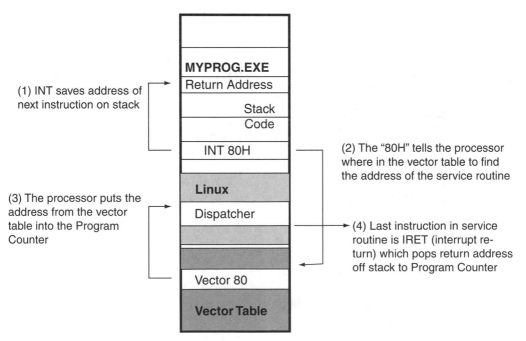

Figure 8-3: Interrupts — How They Work

- The second byte of the INT instruction is an index into the vector table to find the address of the corresponding Interrupt Service Routine (ISR). The processor loads this address into the PC and CS registers and execution proceeds from this point.

- The end of the ISR is indicated by an IRET instruction (Interrupt Return). This pops the PC, CS and PSW off the stack so that execution resumes at the instruction following the INT.

The INT instruction is like the subroutine CALL instruction but with one important difference: Whereas the destination address is embedded in the CALL instruction, with the INT instruction the calling program need not know the address of the ISR! The address is held in the Interrupt Vector Table. Thus, it is ideal for communication between two separately compiled and loaded programs as, for example, an application program and an operating system.

External hardware interrupts operate in a similar fashion as shown in Figure 8-4. A device requiring service asserts an *Interrupt Request (IRQ)* line. When the processor responds with *Interrupt Acknowledge (IAK)*, the device places its interrupt vector number on the data bus. The processor then effectively simulates an INT instruction using the supplied vector index.

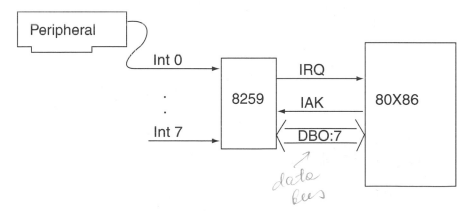

Figure 8-4: Hardware Interrupts

In practice, most systems incorporate a specialized peripheral called an *Interrupt Controller* to manage details like putting the vector index on the data bus at the correct time. The PC architecture includes two 8259 Interrupt Controllers, each able to handle up to eight interrupt inputs. The 8259 provides a mechanism to prioritize interrupts so that more important or critical devices have precedence over less important devices.

Interrupts may also be enabled or disabled. At the processor level, interrupts may be globally enabled or disabled via the STI and CLI instructions. Individual interrupts may be selectively enabled and disabled either at the 8259 Interrupt Controller or at the device itself. In fact the ability to enable and disable interrupts is crucial to the design and implementation of real-time software.

Not surprisingly, asynchronous interrupts are not without their problems. Consider a data acquisition application based on a multi-channel A/D converter as shown in Figure 8-5. Each time the A/D converter takes a set of readings on all channels, it interrupts the processor. The ISR reads the data and stores it in a memory buffer where it is available to the background main program.

Data Acquisition System

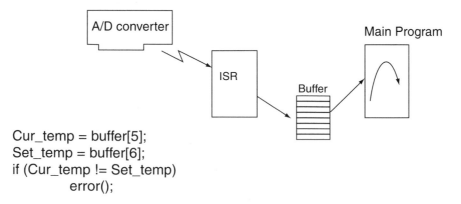

```
Cur_temp = buffer[5];
Set_temp = buffer[6];
if (Cur_temp != Set_temp)
        error();
```

Figure 8-5: Using Interrupts — Example

Interrupt-driven operation allows us to respond to the A/D quickly while the memory buffer effectively "decouples" the background program from the data source—i.e., the background program doesn't need to know where the data comes from.

Consider the code fragment shown in Figure 8-5. Let us say that for testing purposes, we feed the same continuously varying signal into both channels 5 and 6. We would assume that the program would never fail since both channels are measuring the identical signal.

In fact, the program as written is guaranteed to fail because an interrupt can occur between the update of Cur_temp and the update of Set_temp with the result that the value for Cur_temp comes from the *previous* data set while the value for Set_temp comes from the *current* data set. Since the input signal is varying over time and the two data sets are separated by a finite time, the values will be different and the program will fail.

This then is the essence of the real-time programming problem; managing asynchronous interrupts so they don't occur at inopportune times.

There is a simple, albeit inelegant, solution to the problem. We can put a Disable Interrupt (CLI) instruction before the update of Cur_temp and an Enable Interrupt (STI) instruction after the update of Set_temp. This prevents the interrupt from interfering with the variable updates. It turns out that judicious use of STI and CLI is a key element of the "correct" solution but simply scattering STI and CLI throughout the code is like using go to's or global variables. It's just asking for trouble.

Tasks

The "correct" solution is called multitasking, which has proven to be a powerful paradigm for structuring real-time, interrupt-driven systems. I would go so far as to suggest that multitasking is first and foremost a paradigm for safely and reliably handling asynchronous interrupts. The basic idea is that we can break a large problem down into a bunch of smaller, simpler problems. Each one of these sub-problems becomes a task. Each task does *one* thing to keep it simple. Then we pretend that all of these tasks are running in parallel. They aren't really running in parallel unless you have a multi-processor system. On a single processor the tasks share the processor.

Like any program, a task contains code that carries out the function the task is designed to accomplish. This code is embodied in a function that is analogous to the "main" function in a normal C program. What sets a task apart from an ordinary function is that each task has a *context* embodied in its own *stack* (see Figure 8-6).

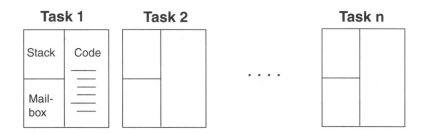

Each task consists of:
- Code to carry out the task's functionality.
- A *stack* to hold the task's "context."
- An optional *mailbox* so the task can communicate with other tasks.

Figure 8-6: What is a Task?

Note, by the way, that it is possible, and sometimes quite useful, to create multiple tasks from the same function. What keeps these tasks separate and distinct is that each one has its own stack. This is really classic object-oriented programming. One could think of the task function as defining a class. Then, each task created from that function is an instance of the class.

Although tasks may be considered to be independent, they typically need to cooperate in order to carry out the overall mission that the system is designed for. Thus, a task requires some form of communication mechanism through which it can communicate and synchronize with other tasks. For the moment, we'll call that mechanism a "mailbox."

Listing 8-2 shows pseudocode of a typical task. The data argument provides a way to parameterize the task in the same manner as argc and argv in main.

This can be especially useful if multiple tasks are derived from the same function. The "uniqueness" of each task is conveyed by the argument value.

A task may start with some initialization (perhaps involving the data argument) after which it usually enters an infinite loop. At some point in the loop, usually near the top, it waits for "something to happen," perhaps the arrival of a message at its mailbox or simply the expiration of some time interval. While it's waiting, the task is not executing, not using the processor. Some other task that's ready to execute is using the processor.

Eventually, the event that the task is waiting for occurs. The task then "wakes up" and, if it received a message, for example, decodes the message and acts on it, often with a large switch statement. After acting on the message, the task returns to wait for something else. Windows programmers will recognize this as the basic Windows programming model.

Note that the reason multitasking works at all is that most tasks spend most of their time waiting for something to happen.

Listing 8-2: Typical task code

```
void task (void *data)
{
    init_task();

    while (TRUE)
    {
      Wait for message at task mailbox();
      switch (message.type)
      {
        case MESSAGE_TYPE_X:
          ...
          break;
        case MESSAGE_TYPE_Y:
          ...
          break;
      }
    }
}
```

Scheduling

Tasks operate under the supervision of the real-time *kernel* that consists of:

- A collection of *services* that implement such things as inter-task communication and synchronization.

- A *scheduler* whose job it is to make sure that the highest priority ready task is the one that's currently executing.

The scheduler treats each task as a state machine. While every kernel has its own, often more complex, state model, Figure 8-7 shows conceptually the minimum state diagram for a task. The states are:

- Running: Only one task, the currently executing task, can be in the Running state. A task can voluntarily transition from Running to Blocked by waiting on an event. In a preemptive system (we're coming to that), the scheduler may cause the Running task to transition to Ready if a higher priority task becomes Ready. This is called *pre-emption*.

- Ready: The task is ready to run but has a lower priority than the currently executing task. The task will transition from Ready to Running when it becomes the highest priority Ready task.

- Blocked: A Blocked task is waiting for some event to occur; a message at a mailbox, a timeout, etc. When the event occurs, the task transitions to Ready.

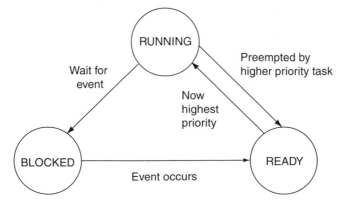

Figure 8-7: Task States

Periodic Scheduling

There are many tasks that simply require waking up periodically, doing something and going back to sleep. There are a couple of approaches to scheduling periodic tasks as shown in Figure 8-8. Every operating system has a function called Delay( ), or some variation thereof, that causes the calling task to be blocked for a specified amount of time, usually expressed in clock ticks. The top half of Figure 8-8 shows what happens when we use Delay( ) to schedule a periodic task, in this case with a period of three clock ticks. The behavior of the system depends on the execution time of the task. If the execution time is less than one clock tick, then the task wakes up every three ticks as desired. However, if the execution time is longer than a clock tick, when the task calls Delay( ) it will still be blocked for three ticks. So in this example the task actually wakes up every fourth tick. That's not what we intended.

An alternative approach, not available with all systems, is to declare the task to be *periodic*. In this case the scheduler wakes the task up at the proper interval regardless of the task's execution time, as shown in the bottom half of Figure 8-8. Instead of calling Delay( ), a periodic task calls a function like WaitTilNext( ) that blocks the task until its next scheduled execution.

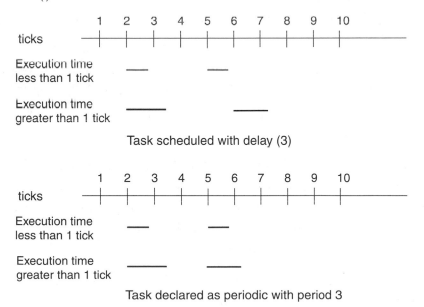

Figure 8-8: Periodic Tasks

Aperiodic Scheduling

Other tasks must respond to events occurring at random times. An event may be the arrival of a network packet, the closure of a switch that indicates a tank is full, or perhaps a conversion is complete on an analog to digital converter and it needs to be read. Very often these asynchronous events are communicated to the computer via interrupts. The interrupt service routine (ISR) must have some way to communicate the occurrence of the interrupt to a task that is responsible for servicing the event. We'll see an example of that in the section on inter-task communication.

Preemptive vs. Non-preemptive Scheduling

There are two fundamental strategies for task scheduling: *preemptive* and *non-preemptive*. Consider two tasks with the lower priority task currently Running and the higher priority task Blocked, waiting for an event that will be signaled by an interrupt.

The upper half of Figure 8-9 shows what happens in the non-preemptive case. The ISR causes the higher priority task to transition to the Ready state but at the end of the ISR, control returns to the lower priority task where it was interrupted. Later when the lower priority task blocks waiting for an event, the higher priority task becomes the Running task.

The lower half of Figure 8-9 shows the pre-emptive case. The difference here is that the scheduler is invoked at the end of the ISR. It determines that the higher priority task is Ready and switches tasks accordingly. The lower priority task is thus *pre-empted*.

A non-preemptive system depends on all tasks being "good citizens" by voluntarily giving up the processor to be sure all tasks get a chance. Early versions of Windows were non-preemptive. Linux is preemptive although standard Linux is not considered real-time due to excessively long periods during which preemption is disabled.

Preemptive systems provide for more predictable response times because a high-priority event is serviced immediately. This is the essence of real-time—

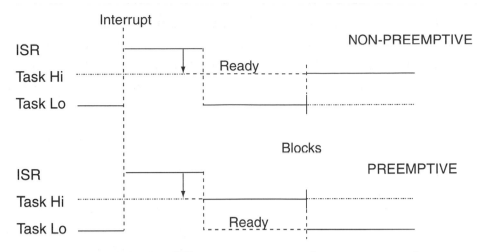

Figure 8-9: Scheduling: Non-Preemptive vs. Preemptive

being able to guarantee the maximum time it takes to respond to an event. In the non-preemptive case there is no guarantee how long it will be before the currently running task gives up the processor. On the other hand, preemptive systems are subject to resource conflict problems that must be carefully considered. We'll see tools for dealing with those problems shortly.

Two other scheduling strategies are employed among tasks of equal priority. In *round robin* scheduling, a task runs until it either blocks waiting for an event or voluntarily yields the processor. The distinction between blocking and yielding is that in the latter case the task is still Ready.

Consider that the Ready List contains three tasks, A, B and C, of equal priority in that order. Task A, at the head of the list, is the Running task. When task A yields, task B becomes the Running task and the Ready List looks like this:

B C A

When task B yields, task C becomes the Running task and the list looks like this:

C A B

Thus, all tasks "get a turn" provided that they all yield. Ready tasks of lower priority do not get to execute until all tasks at this level block.

Timeslicing is a variation on round robin that assigns a maximum time quantum or "slice" to each task to prevent one task from hogging the processor. A task runs until it blocks, yields voluntarily or its timeslice expires. Depending on implementation, the timeslice may be the same for all tasks or each task may get its own timeslice value.

Fundamentally, round robin scheduling is just another form of polling.

Kernel Services

A multitasking kernel is largely defined by the *services* it provides. This set of services constitutes an Application Programming Interface (API) that allows a user to utilize the kernel's features. When describing the functionality of a multitasking kernel, it is useful to present an API to show how the concepts map into real code. For the remainder of this chapter, I'll present a simplified idealized API that expresses the basic functionality. Real implementations will differ in detail from the model presented here but will nevertheless implement the same functionality.

Task API

Let's begin our exploration of the kernel API by examining the services required to manage tasks.

```
task_t *TaskCreate (void (*task)(void *data), void *data, int prior);
status_t TaskStart (task_t *task);
status_t TaskSuspend (task_t *task);
status_t TaskResume (task_t *task);
status_t TaskDelete (task_t *task);
```

Absolutely nothing happens in a multitasking system until we create one or more tasks. To create a task we call a function with a name like TaskCreate. At minimum, we have to give TaskCreate:

- A pointer to the function that implements the task's code (task).

- A pointer to the data that will be passed as the function's argument when it is first called (data).

- The task's priority (prior).

The task create service may return a pointer to a *task control block (TCB)*, identified here as type task_t. This is a data structure containing everything the kernel needs to know about the task. This pointer can then be used as an argument to other task management services.

Note that in this implementation we are assuming that the task create service allocates the TCB *and the stack*. We also assume that the stack size is fixed and is initialized somewhere else. Some implementations require the user to allocate the TCB and/or the stack and stack size. There may be additional arguments like a task name in ASCII or a timeslice value, for example.

The task create service may or may not start the task executing. If it doesn't, a separate task start service is provided. Once a task is executing it may be *suspended*, which simply prevents the task from being scheduled for execution until it is subsequently *resumed*. Finally, a task that is no longer needed may be *deleted*, which removes it from the list of active tasks. In general, task management functions other than TaskCreate return a status_t type indicating whether or not the function succeeded.

Timing API

```
void Delay (unsigned int ticks);
void DelayUntil (time_t *time);
void WaitTilNext (void);
```

Every kernel has a function called something like Delay that blocks the calling task for the specified number of clock ticks. Some systems have a variation on this called DelayUntil that blocks the calling task until a specific time of day. The data type time_t is an unsigned long int representing seconds since midnight, Jan. 1, 1970.

WaitTilNext is only found in systems that support the notion of a periodic task. This function blocks the calling task until the next time the task is scheduled for execution.

Inter-task Communication

Although tasks are considered to be independent, the overall function of the system usually requires that tasks cooperate and communicate with each other. Thus, a key element of any real-time operating system is a set of communication and synchronization services.

There are several communication and synchronization mechanisms in common use:

- Semaphore: Used for synchronization and resource locking.

- Event Flag: Shows that one or more events have occurred. This is an extension of the semaphore that permits synchronizing on a combination of events.

- Mailbox, queue or pipe: Mechanisms for transferring data between tasks

There are many other less widely used mechanisms, such as the ADA "rendezvous" and the "monitor" in Java.

Semaphores

Consider two tasks, each of which wants to print the message "I am Task n" on a single shared printer as shown in Figure 8-10. In the absence of any kind of synchronizing mechanism, the result could be something like "II a amm TaTasskk 12".

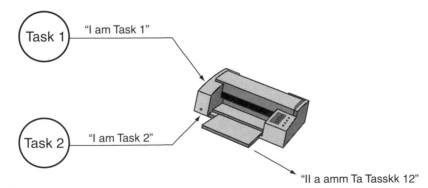

Code in Task n

    printf ("I am Task %d\n", task);

Figure 8-10: Sharing Resources

What is needed is some way to regulate access to the printer so that only one task can use it at a time.

A *semaphore* acts like a key to control access to a resource. Only the task that has the key can use the resource. In order to use the resource (in this case a printer) a task must first *acquire* the key (semaphore) by calling an appropriate kernel service (Figure 8-11). If the key is available, that is the resource (printer) is not currently in use by someone else, the task is allowed to proceed. Following its use of the printer, the task releases the semaphore so another task may use it.

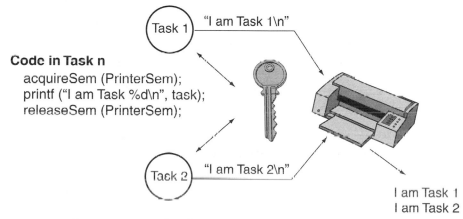

Figure 8-11: Sharing Resources with a Semaphore

If, however, the printer is in use, the task is blocked until the task that currently has the semaphore releases it. Any number of tasks may try to acquire the semaphore while it is in use. All of them will be blocked. The waiting tasks are queued either in order of priority or in the order in which they called acquireSem. The choice of how tasks are queued at the semaphore may be built into the kernel or it may be a configuration option when the semaphore is created.

acquireSem works as follows:

1. Decrement the semaphore value.

2. If the resulting value is greater than or equal to 0, the resource is available and the task can proceed. Otherwise, block the task until another task executes releaseSem.

releaseSem increments the semaphore value. If the resulting value is less than or equal to 0, there is at least one task waiting for the semaphore, so make one of those tasks ready.

In the case of the printer, the semaphore was initialized to 1, reflecting the fact that there is one printer to manage. This is sometimes called a *binary* semaphore to distinguish it from the more general case of the *counting* semaphore, which can be initialized to any non-negative number.

Consider a dynamic memory allocator that manages a fixed number of buffers as shown in Figure 8-12. Here we initialize the semaphore to the number of buffers that are initially available for allocation. When bufReq is called, it first acquires the semaphore, then allocates a buffer. The first ten times bufReq is called, the semaphore is non-negative and the calling task proceeds. The eleventh time, the calling task is blocked until someone else calls bufRel, which releases the semaphore.

Memory Buffer Manager

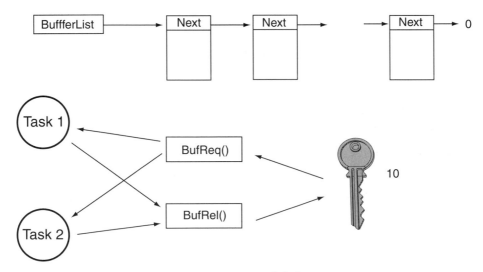

Figure 8-12: Sharing Multiple Resources

Some kernels implement both binary and counting semaphores because in some instances a binary implementation can be more efficient. The binary semaphore is sometimes called a *mutex* meaning "mutual exclusion."

A semaphore can also be used to signal the occurrence of an event as shown in Figure 8-13. For example, how does the system know that an interrupt has occurred? A task that needs to know about the occurrence of an interrupt *pends* on a semaphore. The Interrupt Service Routine services the interrupt and then *posts* to the semaphore. (Note that the terms "pend" and "post" are used more commonly than the terms "acquire" and "release").

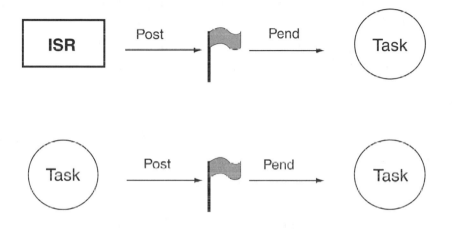

Figure 8-13: Signaling Events Through Semaphores

In the previous examples, the semaphore was initialized to a non-zero value because the resource is initially available. Here it is initialized to 0 so that when the task first pends, it is immediately blocked—the event hasn't occurred yet. When the ISR posts to the semaphore, the task "wakes up" and continues.

When a semaphore serves as a resource lock, many tasks may pend or post to it. However, in the case of signaling or synchronization, the semaphore is typically used exclusively by one ISR and one task.

The same mechanism may be used by one task to signal an event to another task.

Semaphore API

```
semaphore_t *SemCreate (unsigned int value);
status_t SemDelete (semaphore_t *sem);
status_t SemPost (semaphore_t *semaphore);
status_t SemPend (semaphore_t *semaphore, unsigned int timeout);
```

A semaphore object must be created by a service typically called SemCreate that returns a pointer to the data structure, semaphore_t, that represents the semaphore. When creating a semaphore we can give it an initial value that reflects the role that this semaphore is to perform. If the semaphore is intended to protect one or more resources, the initial value is the number of resources that the semaphore is protecting. If the semaphore is intended to signal an event, it is usually given an initial value of zero because the event hasn't yet occurred. A semaphore may be deleted if it is no longer required.

The basic functions of a semaphore are SemPost and SemPend. When posting to a semaphore, the only argument typically required is the identity of the semaphore itself as returned by SemCreate. Pending on a semaphore typically requires two arguments; the identity of the semaphore and an optional timeout argument expressed in system clock ticks. If the timeout argument is non-zero, this represents the maximum time that the task is willing to wait for the semaphore to be posted. If the timeout interval expires SemPend returns an error code.

Mailboxes

In addition to signaling events, tasks often need to share data. This is usually accomplished through the mechanism of a *mailbox*. One task (the sender) posts a *message* to the mailbox while another task (the receiver) pends on the mailbox waiting for a message (Figure 8-14). If no message is posted to the mailbox when the receiving task pends on it, the task is blocked until a message is posted.

In many cases, a message is a pointer. What it points to must be mutually agreed upon by the two tasks involved in the transfer, just as the sender and

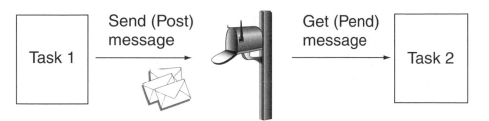

Figure 8-14: Mailboxes and Messages

receiver of a letter must agree on what language to use. Some kernels impose a minimal amount of system structure on messages.

Of course, what the pointer points to must be accessible to both the sender and the receiver. In protected mode systems like Linux this is, by design, not possible. In this case the message is copied from the sender's space to the mailbox and then from the mailbox to the receiver's space.

Depending on implementation, a mailbox may hold only one message or it may be capable of queuing multiple messages that are then delivered to receiving tasks in the order in which they were sent. Some kernels offer the option of sending a "high priority" message that is immediately put at the head of the queue. As with the semaphore, any number of tasks may be waiting at the mailbox queued either in FIFO or priority order.

What happens if a task attempts to send a message to a mailbox that can only hold one message and currently contains a message? There are two implementation-dependent possibilities: the mailbox post service can return an error, or the task can be blocked until it is able to post the message. There's also the possibility that when a task posts a message to a mailbox where no one is waiting, the sender is blocked until another task receives the message.

Mailbox API

```
mailbox_t *MbxCreate (void *message);
status_t MbxDelete (mailbox_t *mbx);
status_t MbxPost (mailbox_t *mbx, void *msg);
void *MbxPend (mailbox_t *mbx, unsigned int timeout, status_t *status);
status_t MbxBroadcast (mailbox_t *mbx, void *msg);
```

The mailbox API is very similar to the semaphore API. Before we do anything else we have to create a mailbox. The mailbox create service, MbxCreate, returns a pointer to a structure of type mailbox_t. MbxCreate may also offer the option of sending an initial message to the mailbox when it is created.

Like the semaphore, the basic operations on a mailbox are pend and post. The arguments to MbxPost are the identity of the mailbox and a pointer to the message. Rather than returning status like most functions, MbxPend generally returns the pointer to the received message and the status variable is passed as an argument.

Some systems include a function like MbxBroadcast that posts the message to all tasks that are currently pending on the mailbox.

Queues and Pipes

```
queue_t *QCreate (int qsize);
status_t QDelete (queue_t *queue);
status_t QPost (queue_t *queue, void *message);
status_t QPostFront (queue_t *queue, void *message);
void *QPend (queue_t *queue, unsigned int timeout,  status_t *status);
status_t QFlush (queue_t *queue);

pipe_t PipeCreate (void);
status_t PipeDelete (pipe_t *pipe);
status_t PipeWrite (pipe_t *pipe, void *buffer, size_t len);
status_t PipeRead (pipe_t *pipe, void *buffer, size_t len);
```

In most cases a queue is simply a mailbox that can hold multiple messages. When creating a queue we usually have to specify its size—that is, the number of messages it can hold. In addition to the normal post and pend calls, a queue will usually have services to post a high-priority message to the front of the queue, QPostFront (), and to flush all messages currently in the queue, QFlush ().

A pipe is a little different. Whereas mailboxes and queues move data in discrete chunks called messages, a pipe is generally a continuous byte stream connecting two tasks. One task reads the pipe, the other writes it. In practice, Unix systems, including Linux, treat pipes as ordinary files using the standard read() and write() functions. The pipe_t is just an array of two long integers where element zero represents the read end of the pipe and element one is the write end. The pipe is created by calling pipe ().

Problems with Solving the Resource Sharing Problem— Priority Inversion

Using semaphores to resolve resource conflicts can lead to subtle performance problems. Consider the scenario illustrated in Figure 8-15. Tasks 1 and 2 each require access to a common resource protected by a semaphore. Task 1 has the highest priority and Task 2 has the lowest. Task 3, which has no need for the resource, has a "middle" priority.

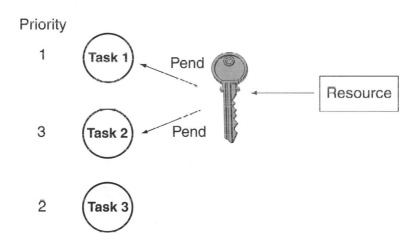

Figure 8-15: Priority Inversion Scenario

Figure 8-16 is an execution timeline of this system. Assume Task 2 is currently executing and pends on the semaphore. The resource is free so Task 2 gets it. Next an interrupt occurs that makes Task 1 ready. Since Task 1 has higher priority, it preempts Task 2 and executes until it pends on the resource semaphore.

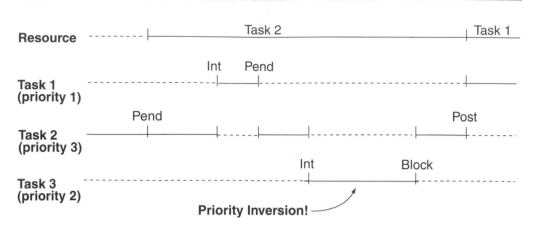

Figure 8-16: Priority Inversion Timeline

Since the resource is held by Task 2, Task 1 blocks and Task 2 regains control. So far everything is working as we would expect. Even though Task 1 has higher priority, it simply has to wait until Task 2 is finished with the resource.

The problem arises if Task 3 should become ready while Task 2 has the resource locked. Task 3 preempts Task 2. This situation is called *priority inversion* because a lower priority task (Task 3) is effectively preventing a higher priority task (Task 1) from executing.

A common solution to this problem is to temporarily raise the priority of Task 2 to that of Task 1 as soon as Task 1 pends on the semaphore. Now Task 2 can't be preempted by anything of lower priority than Task 1. This is called *priority inheritance*. If a kernel makes a distinction between a semaphore and a mutex, the latter will usually incorporate priority inheritance as an optional configuration parameter when the mutex is created.

Another approach, called *priority ceiling*, raises the priority of Task 2 to a specified value higher than that of any task that may pend on the mutex as soon as Task 2 gets the mutex. This is considered to be more efficient because it eliminates unnecessary context switches. No task needing the resource can preempt the task currently holding it.

Interrupts and Exceptions

Interrupt handling in the context of a multitasking kernel requires some special considerations. The problem is that the ISR may need to call a system service to notify a task that something has happened. This in turn could cause the scheduler to be invoked. There are two considerations here:

- The ISR must *not* call any service that would cause a task to block, i.e., MbxPend.

- Generally, the scheduler can't be invoked directly from within an ISR because the system is in a different *context*.

Fundamentally, this means the kernel needs to know when it's running in ISR context rather than task context.

There are two approaches to managing interrupts. I choose to call them the "direct" and "indirect" methods. The top half of Figure 8-17 shows the direct method where the interrupt vector points directly at your Interrupt Service Routine (ISR). You create this connection with a system service like SetVect that places the address of your ISR at the appropriate vector location.

Note the keyword interrupt in the declaration of the interrupt handler function. This turns an ordinary C function into an ISR. It does two things:

- Save all registers on entry to the function and restore them before exiting.

- Replace the normal RET subroutine return instruction with an IRET interrupt return instruction.

Your ISR is responsible for notifying the kernel that the system is operating in interrupt context. The functions IntEnter() and IntExit() provide for this notification.

The indirect method vectors the interrupt into the kernel which recognizes that it's entering interrupt context and performs the actions equivalent to IntEnter(). The kernel then *calls* your handler as an ordinary function. When the handler returns, the kernel performs the equivalent of IntExit() and

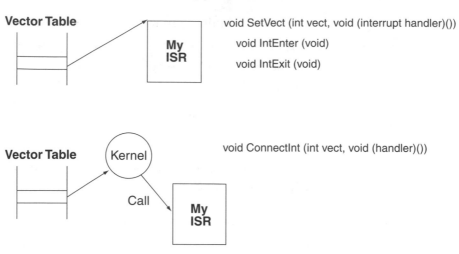

Figure 8-17: Interrupt Management

returns from the interrupt. In this case, the handler needn't do anything with respect to the kernel and indeed doesn't even know it's running in the context of a kernel. Linux uses the indirect method.

Critical Sections

Within the operating system there are sections of code that must execute "atomically"—that is, they must not be interrupted under any circumstance. These are called "critical sections" and must be protected in much the same way as a printer is protected from multiple accesses by a semaphore. Whenever the kernel does a test and set operation on a global variable or updates a global linked list, this is a critical section.

A critical section is typically bracketed by a pair of functions of the form:

    EnterCritical()
    ExitCritical()

Code appearing between these functions is guaranteed not to be interrupted. In practice, these functions are often implemented as in-line assembly language macros that evaluate respectively to:

> Disable interrupts
>
> Enable interrupts

The performance of a real-time operating system is often characterized in terms of its interrupt latency, the maximum time from when an external interrupt is asserted until the ISR begins executing. Interrupt latency is largely determined by the kernel's critical sections. That is, how long does the kernel leave interrupts disabled? Consequently, kernel developers devote considerable effort to keeping the critical sections as short as possible.

It is sometimes useful to implement critical sections in application tasks. Suppose two tasks share access to a global 32-bit counter where one task reads the counter and the other increments it. Incrementing the counter must be done atomically, but on an 8- or 16-bit processor incrementing a 32-bit memory variable requires multiple instructions.

We could use a semaphore to protect the counter, but the overhead of the semaphore operations is probably orders of magnitude greater than the time needed to increment the counter. The more expedient solution is to simply:

```
EnterCritical();
        counter++;
ExitCritical();
```

As long as your critical sections are no longer than the longest critical section in the operating system, then your code is not affecting maximum latency.

Resources

Outside of Linux, the best resource by far for learning about and experimenting with real-time multitasking is MicroC/OS, a small preemptive multitasking kernel written almost entirely in C. MicroC/OS stands for Micro Controller Operating System and is designed to run on small 8- and 16-bit microcontrollers. It is thoroughly described in:

Labrosse, Jean J., *MicroC/OS-II, The Real-time Kernel*, 2ⁿᵈ Ed., 2002, CMP Books.

The book includes a CD with complete source code for MicroC/OS along with several examples that run in a DOS window on a PC. The code is extremely well written and can serve as a model of readability for all programmers. MicroC/OS is used in a number of commercial products and has been certified for use in safety-critical systems under the requirements of RTCA DO-178B.

Other resources related to MicroC/OS are:

www.ucos-ii.com.

Simon, David E., *An Embedded Software Primer*, 1999, Addison-Wesley. David's book uses MicroC/OS as a tool in describing the embedded software development process.

CHAPTER **9**

Linux and Real-time

Why Linux Isn't Real-time

Linux was conceived and built as a general-purpose multiuser operating system in the model of Unix. The goals of a multiuser system are generally in conflict with the goals of real-time operation. General-purpose operating systems are tuned to maximize average throughput even at the expense of latency, while real-time operating systems attempt to minimize, and place an upper bound on, latency, sometimes at the expense of average throughput.

There are several reasons why standard Linux is not suitable for real-time use:

- "Coarse-grained Synchronization" – This is a fancy way of saying that kernel system calls are not preemptible. Once a process enters the kernel, it can't be preempted until it's ready to exit the kernel. If an event occurs while the kernel is executing, the process waiting for that event can't be scheduled until the currently executing process exits the kernel. Some kernel calls, *fork()* for example, can hold off preemption for tens of milliseconds.

- Paging – The process of swapping pages in and out of virtual memory is, for all practical purposes, unbounded. We have no way of knowing how long it will take to get a page off a disk drive and so we simply can't place an upper bound on the time a process may be delayed due to a page fault.

- "Fairness" in Scheduling – Reflecting its Unix heritage as a multi-user time-sharing system, the conventional Linux scheduler does its best to be fair to all processes. Thus, the scheduler may give the processor to a low-priority process that has been waiting a long time even though a higher-priority process is ready to run.

- Request Reordering – Linux reorders I/O requests from multiple processes to make more efficient use of hardware. For example, hard disk block reads from a lower priority process may be given precedence over read requests from a higher priority process in order to minimize disk head movement or improve chances of error recovery.

- "Batching" – Linux will batch operations to make more efficient use of resources. For example, instead of freeing one page at a time when memory gets tight, Linux will run through the list of pages, clearing out as many as possible, delaying the execution of all processes.

You have probably already noticed the consequences of these issues in using Linux or even Windows on your PC. Try moving the mouse while executing a compute-intensive function like rendering a complex graphics image, or while connecting to a dial-up line. The mouse occasionally stops and then jumps because the computer or I/O-bound process has the CPU locked up. In a desktop environment this is nothing more than irritating. In a real-time environment it's unacceptable and may even be catastrophic.

The net effect of all these characteristics is that we can't put an upper bound on the latency that a user task or process may encounter. By definition this is not real-time.

Measuring Latency—an Experiment

Let's try an experiment to see just how much latency varies in standard Linux. Start by untarring the CD file RTAI/Rtdemos.tar.gz into your home directory. You now have a directory named Rtdemos/ and under it is a directory named ProcessJitter/. cd to ProcessJitter/ and open jitter.c.

Move down to around line 60. Jitter uses the select( ) system call to sleep for less than a second, in this case 50 milliseconds. The normal use of select( ) is to wait for any of a specified set of file descriptors to change status. Here we're simply using the timeout argument.

When jitter wakes up it reads the current time of day and computes the interval from the last call to gettimeofday( ). Next it computes, in microseconds, the deviation between the actual time interval and the expected interval of 50 milliseconds. Finally, it updates and prints minimum, maximum, average and current deviations.

Make jitter and start it running. On my lab workstation running X Windows I see a maximum variance in the range of about four to five milliseconds with nothing much going on. The average is on the order of 100 microseconds. Try some simple things like changing directories in a file manager window or opening a text file with the editor. You should see the maximum variance go up.

Now introduce a *real* load on the system. The easiest way to do that is to start up Netscape. You should see maximum variance jump up into the tens of milliseconds. There you have it. Linux is not real-time. To terminate jitter, just hit <Enter>.

Improving Linux Latency

There are some things we can do to improve the latency of standard Linux. Specifically, we can change the kernel's "scheduling policy" and process priority for the jitter process and we can lock the process's memory image into RAM so it won't be paged out.

The default scheduling policy, called SCHED_OTHER, uses a fairness algorithm and gives all processes using this policy priority 0, the lowest priority. This is fine for "normal" processes. The alternate scheduling policies are SCHED_FIFO and SCHED_RR. These are intended for time-critical processes requiring lower latency. Processes using these alternate scheduling policies must have a priority greater than 0. Thus, a process scheduled with SCHED_FIFO or SCHED_RR will preempt any running normal process when it becomes ready.

Conceptually, the kernel maintains a FIFO queue of runnable processes for each possible priority value ranging from 0 to 99. To determine the next process to run, the scheduler finds the highest priority non-empty queue and takes the first entry. If a SCHED_FIFO process is preempted by a higher priority process it remains at the head of its priority queue. When a SCHED_FIFO process becomes runnable after being blocked, it goes at the back of the queue.

A SCHED_FIFO process runs until it blocks or it yields. SCHED_RR is a minor variation on SCHED_FIFO that adds time slicing. If a SCHED_RR process exceeds its time slice, it is placed at the back of its priority queue.

Now look at the section of jitter.c beginning at line 24. Defining the symbol HI_PRIO adds code to set the scheduling policy with a call to sched_setscheduler() and then lock the process in memory with a call to mlockall().

Delete the executable jitter and rebuild it with

> make HI_PRIO=1

You will need to be super user to run this version of jitter. If you're not already root, enter the su command and root password. Now run jitter and start up Netscape again. The maximum variance should be no more than around half of what you saw before. Better, but still not real-time.

Two Approaches

OK, so Linux is not real-time. What do we do about it? Well, there are at least two very different approaches to giving Linux deterministic behavior.

Preemption Improvement

One approach is to modify the Linux kernel itself to make it more responsive. This primarily involves introducing additional preemption points in the kernel to reduce latency. An easy way to do this is to make use of the "spinlock" macros that already exist in the kernel to support symmetric multi-processing (SMP). In an SMP environment spinlocks prevent multiple processors from simultaneously executing a critical section of code. In a uni-processor environment the spinlocks are no ops.

The preemption improvement strategy turns the spinlocks into the equivalent of EnterCritical() and ExitCritical() that we encountered in the last chapter. So whereas the standard kernel prevents preemption unless it's specifically allowed, the preemptible kernel allows preemption unless it's specifically blocked by a critical code section.

Interrupt handling is also modified to allow for rescheduling on return from an interrupt if a higher priority process has been made ready. This approach is often coupled with a new scheduler that provides fixed overhead for real-time tasks. Monta Vista and TimeSys are the principal proponents of preemption improvement.

Note, by the way, that when we speak of kernel preemption, we're referring to *process* latency, not *interrupt* latency. With a standard uni-processor Linux kernel, interrupt latency is on the order of 60 microseconds, depending of course on processor speed. As noted above, maximum process latency for a standard kernel is in the tens of milliseconds. The preemption improvement strategy reduces that to one to two milliseconds.

The advantage to the preemption improvement approach is that the real-time applications run in user space just like any Linux application using the familiar Linux/POSIX APIs. Real-time processes are subject to the same memory protection rules as ordinary processes.

There are a couple of perceived drawbacks:

- Modifying the kernel to this extent is serious business. How can you be sure your mods haven't broken something else? And every time the kernel changes you have to reimplement your modifications.

- It's still not real-time. Latency is reduced but there are simply too many execution paths in the kernel to permit comprehensive analysis of determinism.

For 2.4 series kernels, the preemption improvements are available as a kernel patch. Starting with version 2.5.4-pre6, the preemption patch was merged into the main kernel development tree and is now a configuration option. This effectively nullifies the first objection above. Remember though that 2.5 by definition is a beta release and is not considered stable.

Interrupt Abstraction

In a great many applications, only a small part of the system requires hard real-time determinism. Controlling a high-speed PID loop or moving a robot arm are examples of hard real-time requirements. But logging the temperature the PID loop is trying to maintain, or graphically displaying the current position of the robot arm are generally not real-time requirements.

The alternate, and some would say more expedient, approach to real-time performance in Linux relies on this distinction between what is real-time and what is not. The technique is to run Linux as the lowest priority task (the idle task if you will) under a small real-time kernel. The real-time functions are handled by higher priority tasks running under this kernel. The non-real-time stuff, like graphics, file management and networking, which Linux already does very well, is handled by Linux.

This approach is called "Interrupt Abstraction" because the real-time kernel takes over interrupt handling from Linux. The Linux kernel "thinks" it's disabling interrupts but it really isn't.

Being much smaller and simpler, the real-time OS is amenable to execution time analysis that provides reliable upper bounds on latency. On the other hand, the RTOS introduces its own API and purists insist that this is not "true" Linux. And while this approach also involves modifying the kernel, the extent of the modifications is substantially less than the Preemption Improvement approach.

The real-time tasks run in kernel space[1]. This is a good-news, bad-news situation. The good news is that response times in kernel space are very

---

[1] RTAI does in fact offer real-time functionality in user space. We'll see that in the next chapter.

short. Interrupt response and task switching times under ten microseconds are the norm. The bad news of course is that there's no protection in kernel space and a real-time task can bring down the whole system. The real-time tasks are in effect extensions of the kernel.

There are two major implementations of the Interrupt Abstraction approach:

- RTLinux – This is the original interrupt abstraction implementation. It was developed at the New Mexico Institute of Mining and Technology under the direction of Victor Yodaiken. While an Open Source version of RTLinux is still available, much of the development work is going into a proprietary version called RTLinux/Pro offered by FSM Labs, Inc.

- RTAI – This is an enhancement of RT Linux developed at the Dipartimento di Ingeneria Aerospaziale, Politecnico di Milano under the direction of Prof. Paolo Mantegazza. It is a very active Open Source project with many contributors.

Resources—Obtaining Real-time Linux Implementations

The book CD includes the Open Source version of TimeSys Linux based on kernel version 2.4.7 and RTAI version 24.1.9 that is compatible with all version 2.4 kernels. Here is a short sampling of other real-time Linux resources.

RTLinux is supported by and available from FSM Labs at www.fsmlabs.com/ community.

RTAI is available at www.rtai.org.

The Real Time Linux Foundation is a nonprofit corporation whose charter is to support the real-time Linux community. Visit them at www.realtimelinuxfoundataion.org.

Monta Vista Software offers the preemptible kernel as Monta Vista Linux in both a "Professional Edition" and a high-availability "Carrier Grade Edition" at www.mvista.com.

The TimeSys preemptible kernel is found at www.timesys.com.

LynuxWorks offers a version of RTLinux, licensed from FSM Labs, called BlueCat RT at www.bluecat.com.

## CHAPTER 10

The RTAI Environment

The Real-Time Application Interface, RTAI, uses the Interrupt Abstraction approach to adding deterministic real-time characteristics to the Linux kernel. In this chapter we'll explore the RTAI environment running on the workstation. In principle, RTAI should run on your target with the BlueCat Linux distribution since it's also x86 based. However, BlueCat is based on the version 2.4.10 kernel but RTAI 24.1.9 only has patches back to kernel version 2.4.16. If you really want to try RTAI on your target, go to the RTAI download page and find a version that supports the 2.4.10 kernel.

RTAI has a rather extensive set of APIs. Rather than try to describe all of them in detail, we will focus on some examples that illustrate the basics. The complete APIs are listed in Appendix A.

Installing RTAI

Under the directory /RTAI on the book CDROM are three tar files:

- linux-2.4.18.tar.gz   Untar to /usr/src
- rtai-24.1.9-tar.gz   Untar to /usr/src
- Rtdemos.tar.gz   We untarred this one in the previous chapter

linux-2.4.18.tar.gz creates a directory called linux/ under /usr/src. Note that if there was a symbolic link named linux in /usr/src, it has been replaced by the directory of the same name. You will want to rename linux to linux-

2.4.18-rthal5 to reflect not only the kernel version, but the extensions that we will be adding. Then create the following symbolic links:

```
linux -> linux-2.4.18-rthal5
rtai -> rtai-24.1.9
```

The version numbering scheme used by RTAI reflects the base kernel version that the RTAI version applies to. Thus, the 24 in the RTAI directory name refers to kernel version 2.4.xx.

Patching and Building the Kernel

RTAI requires changes to the base Linux kernel source code. The mechanism for changing released source code in an orderly manner is the *patch* utility. The input to patch is a text file created by the diff utility that compares two files and reports any differences. When an Open Source programmer wants to distribute an upgrade to released source code, she does a diff between the modified code and the original code redirecting the output of diff to a file.

The resulting file is called a "patch file." When you want to implement the modifications on your system, you start with the same original source code and apply the patch. Distributing patch files is generally much more efficient than distributing the entire modified source tree.

Take a look at rtai/patches/. It contains patch files for several kernel versions including ports for ARM, MIPS and Power PC. We'll be using patch-2.4.18-rthal5g to patch the 2.4.18 kernel. Take a look at that file just to get a feel for what a patch file looks like.

In a terminal window, become the Superuser and cd /usr/src/linux. Now execute the command

```
patch –p1 < /usr/src/rtai/patches/patch-2.4.18-rthal5g.
```

patch will list the files it patched.

The –p1 flag tells patch to remove one slash and everything before it from the names of files to be patched. The file names in patch-2.4.18-rthal5g all begin with "linux-2.4.18/". Since we're already in linux-2.4.18-rthal5/, we use –p1 to remove the unnecessary directory information.

Now we need to configure and build our newly patched kernel. Do make xconfig. Open the "Processor type and features" section and note down at the bottom there is now an option for the Real Time Hardware Abstraction Layer together with a note that says this must be yes. Under "Loadable module support" we need to turn off (n) "Set version information on all module symbols". This avoids missing links when loading RTAI modules.

Make sure that other configuration options are correct for your workstation and then build the kernel as described in Chapter 3. Copy the bzImage file to /boot with the name vmlinuz-2.4.18-rthal5. Copy System.map as well. Update lilo.conf or grub.conf as necessary to add the new kernel.

Configuring and Building RTAI

RTAI uses the same configuration mechanism as the kernel. Unfortunately it doesn't support xconfig. cd /usr/src/rtai and do make menuconfig. You'll be asked for the location of the Linux source tree and given a default, which is correct. Eventually the menu in Figure 10-1 will appear. Scroll through the menu to see what the options are. For now we'll stick with the defaults. When you're finished browsing select <Exit>.

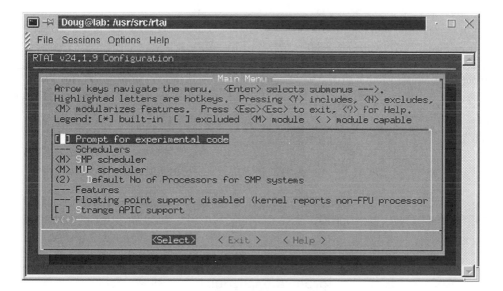

Figure 10-1

To build RTAI, execute the following make commands:

make dep	Builds the dependency files
make	Builds the RTAI modules (takes a while)
make install	Installs the modules in /lib/modules
make dev	Creates inodes in /dev for FIFOs and shared memory
./setsched up	Makes the uni-processor scheduler the default scheduler

When the build is complete, there will be a new directory called /lib/modules/2.4.18-rthal5/rtai containing the RTAI modules. There are quite a few of them but for our purposes the most important are:

rtai	The RTAI interrupt handling functions. Redirects the RTHAL to point to these functions.
rtai_sched	The scheduler. Functions for task management, messaging, semaphores, etc.
rtai_fifos	Real-time FIFOs.
rtai_shm	Shared memory.
rtai_lxrt	User Space real-time tasks.
rtai_pthread	Posix threads API

rtai_sched uses services provided by rtai. In turn, the other modules use services provided by rtai_sched. So there is a definite order to the loading of RTAI modules. rtai must be loaded first, rtai_sched next. The remaining modules don't depend on each other and can be loaded in any order.

As the Superuser, try the following commands (you may need to add /sbin to your path):

```
insmod rtai
insmod rtai_sched
lsmod
cat /proc/rtai/scheduler
rmmod rtai_sched
```

The RTAI Architecture

Figure 10-2 graphically illustrates the basic architecture of RTAI. The real-time kernel effectively intercepts hardware interrupts before they get to the Linux kernel. Linux no longer has direct control over enabling and disabling interrupts. So when Linux says disable interrupts, the RT kernel simply clears an internal software interrupt enable flag but leaves interrupts enabled. When a hardware interrupt occurs, the RT kernel first determines to whom it is directed:

- RT Task – Schedule the task

- Linux – Check the software interrupt flag. If enabled, invoke the appropriate Linux interrupt handler. If disabled, note that the interrupt occurred and deliver it later when Linux re-enables interrupts.

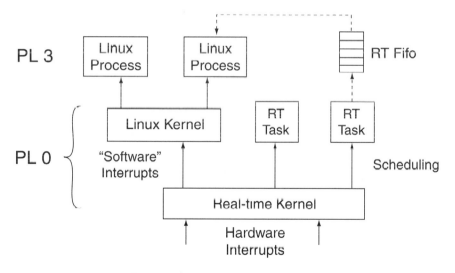

Figure 10-2: RTAI Architecture

The Linux kernel is treated as the lowest priority, or idle, task under the RT kernel and only runs when there are no real-time tasks ready to run.

Of course, there will be times when the RT kernel has to disable hardware interrupts to manage its own critical sections, but these are of much shorter duration than the critical sections in Linux.

The real-time tasks will usually have some need to communicate with user-space processes for things like file access, network communication or user interface. RTAI provides FIFOs and shared memory implementations that provide communication with user space processes.

The Real-Time Hardware Abstraction Layer (RTHAL)

RTAI uses the concept of a *Hardware Abstraction Layer (HAL)* to take over control of interrupt handling from Linux. Listing 10-1 is an excerpt from linux/include/asm-i386/system.h, one of the files modified by the RTAI patch. struct rt_hal is basically a table of function pointers with entries for every function in the Linux kernel that deals in one way or another with interrupts. It also includes some related global variables.

This structure is built into the modified kernel that we just created. When Linux first boots, the rt_hal structure is initialized to point to the original functions in the kernel and Linux behaves normally. Later, when the module rtai.o is installed, the function pointers in rt_hal are changed to point to equivalent functions in the RT kernel. Now the RT kernel is in charge of interrupt handling.

Listing 10-1: RTHAL Structure from system.h

```
struct rt_hal {
        struct desc_struct *idt_table;
        void (*disint) (void);
        void (*enint) (void);
        unsigned int (*getflags) (void);
        void (*setflags) (unsigned int flags);

        void (*mask_and_ack_8259A) (unsigned int irq);
        void (*unmask_8259A_irq) (unsigned int irq);

        void (*ack_APIC_irq) (void);
        void (*mask_IO_APIC_irq) (unsigned int irq);
        void (*unmask_IO_APIC_irq) (unsigned int irq);

        unsigned long *io_apic_irqs;
        void *irq_controller_lock;
        void *irq_desc;
```

```
        int *irq_vector;
        void *irq_2_pin;
        void *ret_from_intr;
        struct desc_struct *gdt_table;
        volatile int *idle_weight;
        void (*lxrt_cli)(void);
} rthal;
```

Why the extra level of indirection—why not simply redefine the functions? Well, suppose you want to restore normal Linux operation. Perhaps the system isn't behaving properly and you'd like to isolate the problem either to Linux or RTAI. Simply removing the rtai module restores normal Linux functionality by pointing the rt_hal entries back to the original kernel functions.

The modifications to the Linux kernel to support the RTHAL are relatively confined and straightforward. Listing 10-2 shows a couple examples comparing the original code to the RTAI modified code. The sti and cli macros get redefined to function calls in the RTHAL. The interrupt management functions simply get "rthal." prepended to them. Altogether about twenty files are involved as listed in Table 10-1.

Listing 10-2: RTHAL Patches

```
                    Linux (system.h)
#define __sti() __asm__ __volatile__ ("sti": : :"memory")
#define __cli() __asm__ __volatile__ ("cli": : :"memory")


                    RTAI (system.h)
#define __sti()        (rthal.enint())
#define __cli()        (rthal.disint())


                    Linux (irq.c)
mask_and_ack_8259A(irq);


                    RTAI (irq.c)
rthal.mask_and_ack_8259A(irq);
```

Table 10-1: Kernel files modified for RTHAL

All paths are relative to /usr/src/linux

Documentation/Configure.help	Add help text for RTHAL option
Makefile	Modify version suffix string
arch/i386/defconfig	Add default for RTHAL option
arch/i386/kernel/entry.S	
arch/i386/kernel/i386_ksyms.c	
arch/i386/kernel/i8259.c	
arch/i386/kernel/io_apic.c	
arch/i386/kernel/irq.c	
arch/i386/kernel/smp.c	
arch/i386/kernel/time.c	
arch/i386/mm/fault.c	
arch/i386/mm/ioremap.c	
include/asm-i386/hw_irq.h	
include/asm-i386/pgalloc.h	
include/asm-i386/system.h	
include/linux/sched.h	
kernel/exit.c	
kernel/fork.c	
kernel/ksyms.c	
kernel/sched.c	
kernel/signal.c	

Intertask Communication and Synchronization

RTAI includes an extensive set of mechanisms to facilitate communication and synchronization among Kernel Space tasks. These include:

- *Semaphores* – Conventional counting semaphores as described in Chapter 8. The semaphore operations are signal and wait.

- *Mailboxes* – RTAI mailboxes are much like the queues described in Chapter 8. The mailbox is created with arbitrary size and arbitrary quantities of data can be written to and read from a mailbox.

- *Messaging* – This is a direct task-to-task communication mechanism. A task can send a single integer directly to another task. The receiving task can wait for a message (integer) from a specific sender or from any task. There is also a full duplex version of messaging called "Remote Procedure Calls."

All of these mechanisms are implemented in the module rtai_sched.

Communicating with Linux Processes

The communication and synchronization mechanisms described above apply only to Kernel Space real-time tasks. RTAI also supports two mechanisms to provide communication between a real-time task and a User Space Linux process. These are RT FIFOs and shared memory.

RT FIFOs

A RT FIFO is a point-to-point link connecting one real-time task to one Linux process. It's very much like a Unix pipe. The implementation allows a FIFO to be bidirectional, but in practice that rarely makes sense. Suppose, for example, that one end of the FIFO writes a command and then immediately tries to read the result of the command it just wrote. Chances are it would just read back the command it wrote. So in practice, FIFOs are unidirectional where the direction is established by the programmer. In the example just cited, you would create two FIFOs—one to send the command and the other to read the response.

User Space processes treat RT FIFOs as character devices, /dev/rtf0 to /dev/rtf63. A process opens a FIFO for reading or writing and then uses read() or write() on the file descriptor to transfer data. The rtf nodes are automatically added to your filesystem by the RTAI build process if you select FIFO support.

Real-time tasks access the FIFO through an RTAI-specific API.

We'll look at the FIFO API in more detail later with a simulated data acquisition example.

Shared Memory

The FIFO model is useful in situations where a relatively small amount of data must be transferred between one RT task and one User Space process more or less synchronously. But there are also situations where multiple processes might require access to data generated by a single real-time task. In this case a shared memory model makes more sense. Remember that we used shared memory back in Chapter 7 for the networked thermostat.

There are also situations where large amounts of data, video frame buffers for example, must be moved quickly between a real-time task in Kernel Space and one or more processes in User Space. Here's another case where shared memory makes sense because the data doesn't have to be copied from one domain to the other. The real-time task writes the shared memory region and the processes read it at their leisure.

Real-time in User Space—LXRT

RTAI and the real-time tasks that it manages run in Kernel Space at Privilege Level 0. As we've seen before, this leads to problems during development because it's difficult to use a source level debugger like GDB on kernel code. Fortunately, RTAI has an escape hatch called LXRT. LXRT allows you to run real-time tasks in User Space using the same API that is provided in Kernel Space RTAI.

How It Works

You create an LXRT task as an ordinary Linux User Space process with a main() function. As part of the initialization, you create a "buddy" task that operates in Kernel Space on behalf of the User Space task. When, for example, your LXRT task calls rt_task_wait_period(), LXRT gets your buddy, running in Kernel Space, to execute the real function. Control returns to the LXRT task only when its buddy wakes up. Similarly you can create RTAI communication and synchronization objects like semaphores and mailboxes. Each of these objects is identified by a name. You pass the object's name to the appropriate initialization or creation function, which returns a pointer to a data structure to be used in accessing the object.

Measuring Latency with LXRT

As an introduction to RTAI we'll run a variation on the latency experiment of the previous chapter. This version is structured as a pair of LXRT tasks. One task, called rt_process, wakes up periodically, reads the current time, computes the deviation from the ideal period and checks min and max values. After some number of passes it sends this information via an RTAI mailbox to the other task, check, which simply displays the results. This is in fact a simple variation on one of the RTAI LXRT examples.

cd Rtdemos/TaskJitter and open rt_process.c. Scroll down into main() around line 69. For LXRT to function correctly, you must use the SCHED_FIFO scheduling policy and lock the process in memory (no paging). The function rt_task_init() creates an LXRT task named "LATCAL" and returns a pointer to a RT_TASK structure. This is a handle by which we'll refer to this task from now on. nam2num() is just a way of "mangling" up to six ASCII characters into a single integer ID. rt_task_init() leaves the task in the suspended state.

Next we create a mailbox named "LATMBX" with a call to rt_mbx_init(), which returns a pointer to a mailbox structure. Within the if statement concerning oneshot, the function start_rt_timer() starts the system timer with the specified period, in this case 100,000 nanoseconds, or 100 microseconds. Note that the argument to start_rt_timer() is in internal timer counts. For the time being, skip over the other code in this if statement as well as the one concerning hardrt. We'll come back to these later.

LATCAL is a periodic task so we start it with a call to rt_task_make_periodic(). The task starts now with the same period as the timer, so it wakes up every timer tick.

Inside the task's while() loop it calls rt_task_wait_period() to block until its next scheduled execution time as specified by rt_task_make_periodic(). Each time the task wakes up, it reads the current time and computes the deviation from the expected wakeup time. After a few thousand passes it sends a data structure to the mailbox with the function rt_mbx_send_if().

The "if" suffix means the message will be sent only if there's space for it in the mailbox. This is a nonblocking version of sending to a mailbox

rt_receive_if() tests to see if the companion task "LATCHK" has sent a message. A non-zero return indicates a message was received. In this case the message value is unimportant; it merely says that LATCHK is going away and so LATCAL should go away too. The function rt_get_adr() looks up a name in the LXRT name space and returns the corresponding handle if the name has been registered by an object initialization function.

Now take a look at check.c. It creates a buddy task for itself and then gets the address (handle) of LATMBX. Note we're assuming that LATCAL starts first. rt_mbx_receive() is a blocking call that doesn't return until it receives a message in the mailbox. Then it simply prints out the various deviation parameters. Finally it tests the flag set by the signal handler to see if it's time to go away. If so, it sends a dummy message to jitter and returns. main() then deletes the task.

Let's try it out. Make the project. In addition to rt_process and check, this makes a third target, jittermod.o, that we'll look at later. You'll need to be Superuser to run the programs because only Superuser processes are allowed to set the scheduler and lock memory.

But before we can run either task, we need to load the LXRT module. Remember that rtai_lxrt depends on other modules. We could just issue individual insmod commands to load the required modules in order. But there's an easier way. Enter the command

```
modprobe rtai_lxrt
```

This checks the dependencies of rtai_lxrt and loads any required modules if they aren't already loaded.

With the RTAI environment set up, we can now run our two tasks. There are two ways to do this. We could start two shell windows and run one program in each window. Or, we can run rt_process as a background task and run check as the foreground task in the same window. Enter

```
./rt_process &
```

The "&" directs bash to spawn a process for rt_process but then immediately come back for another command rather than waiting for rt_process to finish. Now enter

./check

The results are displayed in nanoseconds. Fire up Netscape and watch what it does to the maximum deviation. It's pretty bad, isn't it? In fact, it's not any better than what we saw in the last chapter with straight Linux. So conventional LXRT really is just "soft" real-time. It's useful as a debugging tool before you move your tasks into Kernel Space, but it doesn't buy you anything in terms of deterministic performance.

When you're finished, type <Enter> into the window running check and it will terminate both processes.

Hard Real-time in LXRT

But that's not the end of the story. Later releases of LXRT include a feature that allows you to get hard real-time performance in User Space. The function rt_make_hard_real_time() turns a User Space process into a hard real-time process (perhaps the more accurate term is harder). It does this by playing games with the scheduler and blocking hardware interrupts while a hard real-time process is executing. You can return the process to normal soft real-time by calling rt_make_soft_real_time().

rt_process provides for a couple of run-time parameters, beginning respectively with "h" and "o", to alter its behavior. The "h" stands for hard real-time. If you execute ./rt_process h, the program invokes rt_make_hard_real_time(). Try it out. You should see substantially better results, down in the range of tens of microseconds or less even when starting up Netscape.

One Shot vs. Periodic Timing

RTAI supports two modes of timer tick interrupt, called *periodic* and *one-shot*. In periodic mode you set the timer to interrupt at a specified period. When the timer counter overflows and generates the interrupt, it is automatically reloaded with the correct starting number and starts over. Thus, there is no overhead in servicing the timer chip itself.

In fact this is the way most operating systems work. The downside is that periodic tasks can only be scheduled in increments of the timer tick interval. If the tick interval is one millisecond, for example, the shortest task period is one millisecond. Of course if you need finer granularity you can decrease the tick interval to, say, 100 microseconds, or maybe even 10 microseconds. The problem here is that servicing the timer tick imposes some fixed overhead and the shorter the tick interval, the higher percentage of processor time is devoted to tick servicing. At a tick interval of 10 microseconds, you would probably find the system almost entirely devoted to servicing the tick interrupt, leaving little time to do real work.

RTAI's one-shot timing mode is a solution to this problem. Whereas periodic mode just lets the timer free run, interrupting at the specified interval, one-shot mode reprograms the timer each time it interrupts. That is to say, at every tick interrupt, we compute the time to the next "event" and program the timer for that interval. The resolution of the timing interval is now determined by the clock driving the timer and not the periodic interval of timer interrupts.

Consider a simple example involving the following three periodic tasks:

Task	Interval
Task1	1.3 milliseconds
Task2	600 microseconds
Task3	2.1 milliseconds

Assume for the sake of simplicity that all three tasks are started simultaneously. They will be ordered on a waiting list with the shortest interval first and the other tasks expressed as time remaining after the first interval expires:

Task	Remaining Interval (microseconds)
Task2	600
Task1	700
Task3	1500

The timer is set for 600 microseconds. When the interrupt occurs, Task2 is made ready and the list is updated to reflect the next interval. In this case,

Task2's 600-microsecond period is still less than the remaining intervals for the other tasks and so the timer is again set for 600 microseconds and the list is updated as follows:

Task	Remaining Interval (microseconds)
Task2	600
Task1	100
Task3	900

At the next timer interrupt, Task2 is again made ready, the list is updated as follows and the timer is set for 100 microseconds:

Task	Remaining Interval (microseconds)
Task1	100
Task2	500
Task3	800

The tradeoff for this flexibility is more timer overhead due to the need to recalculate remaining intervals and reprogram the timer at each tick interrupt. To give a feel for the overhead, the RTAI group claims that on a 233-MHz Pentium III, periodic mode supports tick rates up to 90 kHz while one-shot mode supports up to about 30 kHz.

One-shot mode is invoked by calling rt_set_one_shot_mode(). There is also a function rt_set_periodic_mode(), which happens to be the default.

The "o" run-time parameter for rt_process sets up one-shot timing mode. Take a look at the source code of rt_process down around line 85. When one-shot mode is invoked the argument to start_rt_timer() is not used. In principle, the translation from nanoseconds to counts could be different in one-shot mode. In practice, on Intel processors, it's the same. Note that we could pull the two lines,

```
period = nano2count (PERIOD);
start_rt_timer (period);
```

outside the if statement. But it was done this way to make the above distinctions clearer.

Try it out and see if there are any noticeable differences in jitter between periodic and one-shot timer modes.

Moving to Kernel Space

LXRT is a good way to get started with RTAI because you can use DDD to see what's going on. But now it's time to see what happens in Kernel Space. Take a look at rt_module.c in TaskJitter/. Go down to the init_module() function around line 83. The first thing to notice is that rt_task_init() takes more arguments than the LXRT form. In this case we have to specify a function, latency(), that implements the task as well as the size of the task's stack. On the other hand, we don't need to give the task a name. Note also that oneshot is now a module parameter.

init_module() does basically the same initialization as rt_process.c. Likewise cleanup_module() does the same basic post processing as rt_process.c. Look at the function latency() starting at line 33. It first registers the mailbox name so that the check task running in User Space can find it. Then it enters almost the exact same infinite loop as rt_process. The only difference is that it doesn't test for a message from the check task. The only sensible way to stop a Kernel Space task is to remove the module.

Note the use of rt_printk() inside latency(). printk() itself is not safe to use inside an RTAI task. printk() "thinks" it has disabled interrupts but of course it has only disabled Linux interrupts. The "real" interrupts are under control of RTAI. rt_printk() manages interrupts at the RTAI level and thus is safe to use within RT tasks.

Try it out. On my system the numbers all come back zero, even while loading Netscape. Hmmm... That's enough to make you suspicious that something's wrong here. Check the code. I think you'll find that the algorithm inside the loop in latency() is identical to the loop in rt_process.c. I even tried setting samp.min and samp.max to non-zero values before the call to rt_mbx_receive() just to be sure the data structure was being written from the mailbox. It is. We've achieved genuine hard real-time performance.

RTAI /proc Files

Upon installing the rtai module, you'll find a new subdirectory under /proc, /proc/rtai. Each RTAI module creates its own file in /proc/rtai to convey its own status information. Just for kicks, take a look at /proc/rtai/scheduler both before and after installing rtai_lxrt. The scheduler file lists the real-time tasks and useful information about them. After installing rtai_lxrt we find 16 tasks in the list! Look at /proc/rtai/lxrt and you'll find that it too has created 16 tasks. If you start rt_process and look at /proc/rtai/lxrt again, you'll find two more objects in the list, the LATCAL task and LATMBX mailbox.

Real-time FIFOs and Shared Memory

cd $(HOME/)Rtdemos/FIFO and open the file data_acq.c. This is a pseudo data acquisition application that uses an RTAI FIFO and a shared memory region to communicate with a logger process. data_acq supports multiple channels and we can control the acquisition process through a channel t data structure defined in data_acq.h. We can, for example, set the "sample rate" and "gain" independently for each channel. In this pseudo version data_acq simply generates a sawtooth waveform on each channel at the specified rate and with a specified range.

Go to the function init_module() at line 89. The first thing it does is to create a FIFO for transferring data from the RT task to the logger. The FIFO is identified by an integer and we arbitrarily set the size to 1024 bytes. Next we create a signal handler for the FIFO. This function is a callback that is invoked whenever the User Space end of the FIFO is accessed.

It's quite possible for the data FIFO to fill up, especially if the data_acq task is started before the logger. In that case data_acq can pend on a semaphore until the logger reads something out of the FIFO, thereby making room available again. The FIFO signal handler posts the semaphore when it is invoked by the data FIFO being read and if the data_acq task is pended. init_module() creates the semaphore.

Next we allocate a shared memory region for the channel_t control structures. rtai_kmalloc() is the RTAI-safe version of kmalloc(). RTAI shared memory uses the same object name space as LXRT to allow User Space processes to access the shared memory. Finally init_module() initializes and starts the data_acq task as we've seen before.

The data_acq task begins at line 45. By convention, a channel is not enabled for acquisition if its sample_period field is 0. All channels are initialized to disabled. The logger will set appropriate operational values for each channel.

In the main loop, each time the task wakes up it loops through all the channels in the channel[] array to find any that require a data sample to be generated. For each channel whose sample period has expired, it fills in a data_point_t structure, including channel number, data value and time stamp, and sends this to the data FIFO. The function value of rtf_put() is the number of bytes written, which may be less than the number requested if the FIFO fills up. In this case data_acq pends on the semaphore.

If in fact data_acq writes less than a full data_point_t record to the FIFO, it might be possible for the logger to get out of sync by reading a partial record. It works out in this instance because the data_point_t record is 16 bytes, an integral sub-multiple of the FIFO size. So either a full record will be written or nothing will be written. Likewise, on the logger's end either a full record is read or the FIFO is empty. Note also that when the FIFO fills up, the current data point is simply thrown away. Is that good policy? Well, if the FIFO fills up, the task blocks and we'll probably lose data points anyway. Better that the logger be able to keep up.

Now take a look at logger.c. It starts out with an array of channel_t structures that initializes four channels. A signal handler intercepts Control-C to terminate the program gracefully. Logger writes the data it receives from data_acq to a disk file whose name is passed as a run-time parameter. Note how the FIFO is opened as an ordinary file for read-only access.

Logger uses the same function as data_acq, rtai_malloc() to attach to the shared memory region. The first call to rtai_malloc() for a given region name causes it to be allocated. Subsequent calls simply make the connection. After copying the channel_t data structures to shared memory, logger just reads the data FIFO and writes to disk until it's terminated.

Try it out. After making the targets, you'll need to insmod the following modules if they're not already installed:

> rtai
> rtai_sched
> rtai_fifos
> rtai_shm
> data_acq.o.

Interestingly enough, rtai_shm generates the warning message about "tainting" the kernel because it doesn't specify an Open Source license (remember the "hello" module back in Chapter 6?). Then run logger. Let it run for maybe ten seconds to acquire some data, and then type Control-C to stop it. Use the following command to examine the file created by logger and verify that it really did log data.

> od –t xl *datafile* > *datafile*.txt

where *datafile* is the filename you passed to logger. od translates a binary file into a readable format (octal by default) and dumps it to stdout. –t specifies the output format, in this case hexadecimal (x) integers (I).

Suggested Exercises

This has been a fairly brief tour of the major features of RTAI. Having gone through the data acquisition example, here are some suggested projects for exploring further.

- Create one or more simple utilities to control the data acquisition parameters for a specified channel such as sample period and range or gain. A utility to return the number of samples acquired on a given channel would also be useful.

- Port the thermostat example to RTAI. Provide a utility to adjust the setpoint through shared memory. The current temperature could either be sent through a FIFO or also put in shared memory.

- Incorporate the parallel port hardware and device driver of Chapter 6 into the RTAI thermostat. Remember that RTAI tasks run in Kernel Space so they can directly access I/O ports. In fact, you could think of RTAI as a sort of "super device driver" model.

11

Posix Threads

One of the objections to the RTAI approach to adding real-time functionality to Linux is that it introduces a new, nonportable API to support the real-time kernel. One solution to this problem is to implement the Posix threads API, also known as Pthreads, as a wrapper on top of the native RTAI API.

Posix, also written POSIX, is an acronym that means Portable Operating System Interface with an X thrown in for good measure. POSIX represents a collection of standards defining various aspects of a portable operating system based on UNIX. These standards are maintained jointly by the Institute of Electrical and Electronic Engineers (IEEE) and the International Standards Organization (ISO). Recently the various documents have been pulled together into a single standard in a collaborative effort between the IEEE and The Open Group (see the Resources section below).

In general, Linux conforms to Posix. The command shell, utilities and system interfaces have all been upgraded over the years to meet Posix requirements. But in the context of real-time multitasking, as in RTAI, we are specifically interested in the Posix Threads interface known as 1003.1c.

There is another advantage to using the Pthreads API. Linux itself implements Pthreads in User Space. In effect Pthreads becomes a portable alternative to LXRT. You can develop your real-time application in User Space, using DDD/GDB for debugging, and then port it with relatively little change to Kernel Space RTAI.

Unfortunately, the RTAI Pthreads implementation isn't quite complete and some aspects of Pthreads behavior don't map directly into the RTAI tasking model. I'll describe the Pthreads API in general and point out those areas where the RTAI implementation doesn't quite match.

The header file that prototypes the Pthreads API is pthread.h and resides in the usual directory for library header files, /usr/include.

Threads

Fundamentally a thread is the same thing we were calling a task in Chapter 8. It is an independent thread of execution embodied in a function. The thread has its own stack.

```
int pthread_create (pthread_t *thread, pthread_attr_t *attr, void *(* start_ routine)
    (void *), void *arg);
void pthread_exit (void *retval);
int pthread_join (pthread_t thread, void **thread_return);
pthread_t pthread_self (void);
int sched_yield (void);
```

The mechanism for creating and managing a thread is analogous to creating and managing the tasks. A thread is created by calling pthread_create() with the following arguments:

- pthread_t – A *thread object* that represents or identifies the thread. pthread_create() initializes this as necessary.

- Pointer to a thread *attribute* object – Often it is NULL. More on this later.

- Pointer to the start routine – The start routine takes a single pointer to void argument and returns a pointer to void.

- Argument to be passed to the start routine when it is called.

A thread may terminate by calling pthread_exit() or simply returning from its start function. The argument to pthread_exit() is the start function's return value.

In much the same way that a parent process can wait for a child to complete by calling waitpid( ), a thread can wait for another thread to complete by calling pthread_join( ). The arguments to pthread_join( ) are the ID of the thread to wait on and a place to store the thread's return value. The calling thread is blocked until the target thread terminates. RTAI Pthreads doesn't implement pthread_join( ) as the tasking model doesn't support the notion of joining.

A thread can determine its own ID by calling pthread_self( ). Finally, a thread can voluntarily yield the processor by calling sched_yield( ).

Note that most of the functions above return an int value. This reflects the Pthreads approach to error handling. Rather than reporting errors in the global variable errno, Pthreads functions report errors through their return value. Appendix B gives a more complete description of the Pthreads API including a list of all error codes.

Thread Attributes

POSIX provides an open-ended mechanism for extending the API through the use of *attribute objects*. For each type of Pthreads object there is a corresponding attribute object. This attribute object is effectively an extended argument list to the related object create or initialize function. A pointer to an attribute object is always the second argument to a create function. If this argument is NULL, the create function uses appropriate default values. This also has the effect of keeping the create functions relatively simple by leaving out a lot of arguments that normally take default values.

An important philosophical point is that all Pthreads objects are considered to be "opaque." This means that you never directly access members of the object itself. All access is through API functions that get and set the member arguments of the object. This allows new arguments to be added to a Pthreads object type by simply defining a corresponding pair of get and set functions for the API. In simple implementations the get and set functions may be a pair of macros that access the corresponding member of the attribute data structure.

```
int pthread_attr_init (pthread_attr_t *attr);
int pthread_attr_destroy (pthread_attr_t *attr);
int pthread_attr_getdetachstate (pthread_attr_t *attr, int *detachstate);
int pthread_attr_setdetachstate  (pthread_attr_t *attr, int detachstate);

Scheduling Policy Attributes
int pthread_attr_setschedparam (pthread_attr_t *attr, const struct sched_param *param);
int pthread_attr_getschedparam (const pthread_attr_t *attr, struct sched_param *param);
int pthread_attr_setschedpolicy (pthread_attr_t *attr, int policy);
int pthread_attr_getschedpolicy (const pthread_attr_t *attr, int *policy);
int pthread_attr_setinheritsched (pthread_attr_t *attr, int inherit);
int pthread_attr_getinheritsched (const pthread_attr_t *attr, int *inherit
```

Before an attribute object can be used, it must be initialized. Then any of the attributes defined for that object may be set or retrieved with the appropriate functions. This must be done before the attribute object is used in a call to pthread_create(). If necessary, an attribute object can also be "destroyed." Note that a single attribute object can be used in the creation of multiple threads.

The only required attribute for thread objects is the "detach state." This determines whether or not a thread can be joined when it terminates. The default detach state is PTHREAD_CREATE_JOINABLE meaning that the thread can be joined on termination. The alternative is PTHREAD_CREATE_DETACHED, which means the thread can't be joined.

Joining is only useful if you need the thread's return value. Otherwise it's better to create the thread detached. The resources of a joinable thread can't be recovered until another thread joins it, whereas a detached thread's resources can be recovered as soon as it terminates. In RTAI Pthreads the only valid value for detach state is PTHREAD_CREATE_DETACHED and pthread_attr_setdetachstate() doesn't really do anything.

There are also a number of optional scheduling policy attributes that RTAI Pthreads implements. See Appendix B for more details on these.

Synchronization—Mutexes

Pthreads uses the mutex as its primary synchronization mechanism.

```
pthread_mutex_t mutex = PTHREAD_MUTEX_INITIALIZER;

int pthread_mutex_init (pthread_mutex_t *mutex, const pthread_mutexattr_t *mutex_attr);
int pthread_mutex_destroy (pthread_mutex_t *mutex);

int pthread_mutex_lock (pthread_mutex_t *mutex);
int pthread_mutex_unlock (pthread_mutex_t *mutex);
int pthread_mutex_trylock (pthread_mutex_t *mutex);
```

The Pthreads mutex API follows much the same pattern as the thread API. There is a pair of functions to initialize and destroy mutex objects and a set of functions to act on the mutex objects. This slide also shows an alternate way to initialize statically allocated mutex objects. PTHREAD_MUTEX_INITIALIZER provides the same default values as pthread_mutex_init().

Two operations may be performed on a mutex: *lock* and *unlock*. The lock operation causes the calling thread to block if the mutex is not available. *trylock* allows you to test the state of a mutex without blocking. If the mutex is available *trylock* returns success and locks the mutex. If the mutex is not available it returns EBUSY.

Mutex Attributes

```
int pthread_mutexattr_init (pthread_mutexattr_t *attr);
int pthread_mutexattr_destroy (pthread_mutexattr_t *attr);

int pthread_mutexattr_getkind_np (pthread_mutexattr_t *attr, int *kind);
int pthread_mutexattr_setkind_np (pthread_mutexattr_t *attr, int kind);

kind =  PTHREAD_MUTEX_FAST_NP
        PTHREAD_MUTEX_RECURSIVE_NP
        PTHREAD_MUTEX_ERRORCHECK_NP

int pthread_mutexattr_getprioceiling (const pthread_mutexattr_t *mutex_attr, int
        *prioceiling);
int pthread_mutexattr_setprioceiling (pthread_mutexattr_t *mutex_attr, int prioceiling);
int pthread_mutexattr_getprotocol (const pthread_mutexattr_t *mutex_attr, int *protocol);
int pthread_mutexattr_setprotocol (pthread_mutexattr_t *mutex_attr, int protocol);

protocol =  PTHREAD_PRIO_NONE
            PTHREAD_PRIO_INHERIT
            PTHREAD_PRIO_PROTECT
```

Mutex attributes follow the same basic pattern as thread attributes. There is a pair of functions to create and destroy a mutex attribute object. The only mutex attribute in RTAI Pthreads is a Linux-specific nonportable extension called "kind." The Pthreads standard explicitly allows nonportable extensions. The only requirement is that any symbol that is nonportable have "_np" appended to its name, as shown here.

What happens if a thread should attempt to lock a mutex that it has already locked? Normally the thread would simply hang up. Linux offers a way out of this dilemma. The "kind" attribute alters the behavior of a mutex when a thread attempts to lock a mutex that it has already locked:

Fast – This is the default type. If a thread attempts to lock a mutex it already holds it is blocked and thus effectively deadlocked. The fast mutex does no consistency or sanity checking and thus it is the fastest implementation.

Recursive – A recursive mutex allows a thread to successfully lock a mutex multiple times. It counts the number of times the mutex was locked and requires the same number of calls to the unlock function before the mutex goes to the unlocked state.

Error checking – If a thread attempts to recursively lock an error-checking mutex, the lock function returns immediately with the error code EDEADLK. Furthermore, the unlock function returns an error if it is called by a thread other than the current owner of the mutex.

Note the "_NP" in the constant names.

Optionally, a Pthreads mutex can implement the priority inheritance or priority ceiling protocols to avoid priority inversion as discussed in Chapter 8. The mutex attribute *protocol* can be set to "none," "priority inheritance" or "priority ceiling." The *prioceiling* attribute sets the value for the priority ceiling. The protocol and prioceiling attributes are not available in RTAI Pthreads.

Communication—Condition Variables

There are many situations where one thread needs to notify another thread about a change in status to a shared resource protected by a mutex. Consider the situation in Figure 11-1 where two threads share access to a queue. Thread 1 reads the queue and Thread 2 writes it. Clearly each thread requires exclusive access to the queue and so we protect it with a mutex.

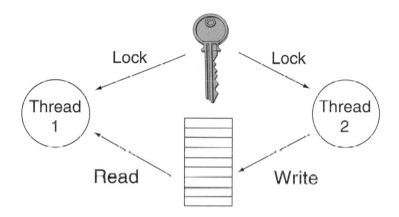

Figure 11-1: Condition Variable

Thread 1 will lock the mutex and then see if the queue has any data. If it does, Thread 1 reads the data and unlocks the mutex. However, if the queue is empty, Thread 1 needs to block somewhere until Thread 2 writes some data. Thread 1 must unlock the mutex before blocking or else Thread 2 would not be able to write. But there's a gap between the time Thread 1 unlocks the mutex and blocks. During that time, Thread 2 may execute and not recognize that anyone is blocking on the queue.

The condition variable solves this problem by waiting (blocking) with the mutex locked. Internally, the conditional wait function unlocks the mutex, allowing Thread 2 to proceed. When the condition wait returns, the mutex is again locked.

```
pthread_cond_t cond = PTHREAD_COND_INITIALIZER;

int pthread_cond_init (pthread_cond_t *cond, const pthread_condattr_t *cond_attr);
int pthread_cond_destroy (pthread_cond_t *cond);

int pthread_cond_wait (pthread_cond_t *cond, pthread_mutex_t *mutex);
int pthread_cond_timedwait (pthread_cond_t *cond, pthread_mutex_t *mutex,
        const struct timespec *abstime);
int pthread_cond_signal (pthread_cond_t *cond);
int pthread_cond_broadcast (pthread_cond_t *cond);
```

The basic operations on a condition variable are *signal* and *wait*. Signal wakes up one of the threads waiting on the condition. The order in which threads wake up is a function of scheduling policy. A thread may also execute a *timed wait* such that if the specified time interval expires before the condition is signaled, the wait returns with an error. A thread may also *broadcast* a condition. This wakes up all threads waiting on the condition.

Condition Variable Attributes

Pthreads does not define any required attributes for condition variables although there is at least one optional attribute. RTAI Pthreads provides the functions to initialize and destroy a condition variable attribute object but does not implement the optional attribute.

Pthreads in User Space

As an exercise in using Pthreads in User Space, we'll recast the data_acq example from the last chapter. This is again a four-channel simulated data acquisition application where each channel generates a simple sawtooth of a specified amplitude. The acquired data is displayed on the screen rather than written to a file. The primary difference is that rather than have a single task manage all four channels, each channel now gets its own thread. The motivation behind this approach is that a thread that only manages one channel is probably simpler than a thread that manages "n" channels.

Each of the "channel" threads writes its data into a common data structure protected by a condition variable. A single "display" thread reads the structure and writes the data to the screen. Finally, a "command" thread accepts simple operator commands from the keyboard and updates the channels accordingly. The system architecture is graphically illustrated in Figure 11-2.

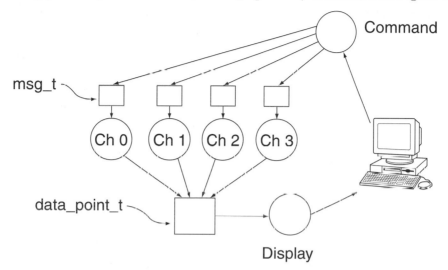

Figure 11-2: data_acq in User Space

cd /$HOME/Rtdemos/Posix and open data_acq.c. Near the top at line 21 is an array of channel_t data structures. Looking at data_acq.h, you'll see that many of the fields in channel_t have been removed. That's because they can now be local variables inside the data_acq() thread. Other fields have been added to support Pthreads. Note that the data_point_t typedef includes an #ifdef to specify an alternate definition of timestamp. It turns out that we need to deal with time a little differently in Kernel Space.

Next is a data_point_t structure for communicating data from the channel threads to the display thread. This structure requires a mutex and a condition variable to protect it.

Before looking at the data_acq() thread, let's move down to main() at line 166. It starts by initializing the four data acquisition channels using the channel[] array. For each channel it creates a mutex, a condition variable and a thread. The argument passed to the thread is the channel's channel_t structure.

Next we initialize the display. We're using the Curses library as we did in devices.c back in Chapter 5. Then we create the Command() thread that will monitor the keyboard for operator input. Finally, main() calls Display(), which becomes the main thread that waits for data from the channel threads and displays it. We could have just left the display loop in main() but when it's time to move to Kernel Space it will be useful to have a separate function.

Move up to Display() at line 146. Here, then, is our first example of using mutexes and condition variables. The display thread locks the mutex associated with the display data structure (m_display) and waits on the corresponding condition variable (c_display). pthread_cond_wait() won't return until someone else, one of the channel threads, signals that it has put something in the display structure. Upon waking up, the display thread copies the data to a local variable before unlocking the mutex.

Now let's go back and look at data_acq() starting at line 26. We first copy a number of fields from the channel_t structure passed as an argument. This is as much for readability as anything else. We maintain the same convention that a sample period of zero means the channel is disabled. data_acq() is now ready to go to sleep and here's where it gets interesting. There are two different circumstances in which data_acq() should wake up:

- It's sampling and the sample period has expired
- It received a message from Command() to alter one of its sampling parameters.

data_acq() receives messages from Command() through a buffer contained in its channel_t structure. Since this buffer is accessed by two threads, it must be protected by a mutex. We lock the associated mutex in the channel_t structure before accessing the message buffer. But there's no point in trying to read the message buffer until the Command() thread sends us something. So we'll wait on the condition variable in channel_t and let Command() signal us when something changes.

If the channel is not sampling (sample_period == 0), then we can just call pthread_cond_wait() until the operator decides to turn this channel on. If the channel is sampling, we could call some "sleep" function to just delay for the sample period but at the same time, we also have to respond to operator commands. Here's where pthread_cond_timedwait() is useful. We can call pthread_cond_timedwait() with the sample period as the timeout value. When it returns, the status value will indicate whether the call timed out (status == ETIMEOUT) or somebody signaled us (status == 0).

Upon waking up from pthread_cond_timedwait(), we have to compute the next wakeup time. That's because timed wait takes absolute time as its timeout argument[1]. So we add sample_period, converted to nanoseconds, to the timeout argument.

If status indicates a timeout, we're not interested in the message buffer so we can immediately unlock the mutex. Now we need exclusive access to the display data structure so we lock its mutex. Then we copy the relevant information to the display structure, signal its condition variable and unlock the mutex.

If status is zero, Command() has put something into the message buffer. In this case we leave the mutex locked while we deal with the message. The major complication here is that if the message changes the sample period, we have to compute a new wakeup time.

Let's move on to the Command() thread at line 108. The getstr() function blocks the thread until the operator types <Enter>. Upon returning, we parse the command string. The command syntax is similar to what you did in Chapter 5 adding programmable parameters to the thermostat. There are four valid commands:

- 'c' – channel. The Command() thread handles this internally. Subsequent commands are directed to this channel until the channel value is changed.

---

[1] I find this somewhat "klutzy." Other systems I'm familiar with tend to treat timeouts as intervals rather than absolute time. You just call the function again with the same timeout interval. But more than klutzy, this is just the sort of thing we don't need in embedded systems. Y2K was not the disaster that the popular press predicted precisely because the vast majority of embedded systems have no concept of absolute time. Fortunately, Unix absolute time won't overflow until Feb. 7, 2106, but if you happen to treat time as a signed value, it overflows on Jan. 19, 2038. Hmmm…

- 'q' – quit. Stop everything. The call to exit() cancels all threads. This is probably not the cleanest way to stop the system. The clean solution would be for Commnad() to notify the other threads that it's time to go away and have them commit suicide.

- 'r' – range. The maximum value for data generated by this channel.

- 's' – sample period. Set the sample period for the current channel in milliseconds.

In the interest of expediency, Command() does no error or sanity checking. Except for 'c' and 'q' (case sensitive, by the way), all perceived command tokens are passed on to the current channel. There's no check that a channel value is in range.

Keyboard input is not echoed. This would require sharing the display between two threads so that Command() could move the cursor to a command line field and then restore it as each character is typed. This in turn requires that the display be protected by a mutex.

Making data_acq

Have a look at the Makefile. Ignore the commented section; we'll come back to that later. The logger target also isn't relevant yet. Two libraries have been added to the compile command for data_acq with the "-l" option. The names are fairly obvious. The pthread library contains the Pthreads functions and curses is the curses library.

There's also a compile time symbol, _REENTRANT. The original Linux library routines assumed that only a single thread of execution would be running in a process. An example of this assumption is the global variable errno. That works fine if there's only a single execution thread. But when multiple threads are executing "simultaneously," it would be quite possible for a function called by one thread to set errno only to have the value changed by a function called by a second thread before the first thread read the value intended for it. That's why Pthreads functions return their error codes as the function value rather than using errno.

But it goes deeper than that. Functions that could be called simultaneously by more than one thread must be *reentrant*. A function is reentrant if it doesn't use any statically allocated resources like global memory. C functions tend to be naturally reentrant provided they only reference local variables, because these variables are allocated on the stack. Each thread gets its own copy of the local variables. But some library functions, the printf() family for example, use statically allocated buffers. The purpose of the _REENTRANT flag then is to alert the compiler to substitute reentrant versions of these functions that are "thread safe."

So try it out. make data_acq and run it. Enter the command "c 0 s 500" to start channel 0 sampling. Then enter similar commands for the other channels.

Debugging Multi-threaded Programs

The nice thing about running Pthreads in User Space is that you can debug your program with DDD/GDB. Multi-threading does introduce some complications to the debugging process but, fortunately, GDB has facilities for dealing with those. Run data_acq under DDD and set the initial breakpoint at the line:

```
channel[i].number = i;
```

near the beginning of main(). But before running the program, execute View-> Execution Window to bring up a separate window for the program to write its output. Curses output doesn't work very well in the gdb console window.

Now run the program and when it stops at the breakpoint you'll see the following messages in the gdb console window:

```
[New Thread 1024 (runnable)]
[Switching to Thread 1024 (runnable)]
```

GDB has recognized that it's running a multi-threaded environment. Let the program continue. When it stops at that breakpoint again, you'll see two more "New Thread" messages with different numbers. One of those is the channel thread created by the "for" loop, the other is created by Pthreads for its own use. Let the program continue again and another New Thread message will appear.

Delete the breakpoint and let the program continue to the while (1) statement at the top of the display loop. At this point all the threads have been created. Execute Status->Threads. You'll see a list of all the threads and their current execution point. The currently executing thread is highlighted. Note that four of the threads, the channels, are all at the same execution point.

Execute Status->Backtrace to bring up the call stack for the current thread. This shows that we're in main() and that main() was called from __libc_start_main(). Select one of the threads that's at sigsuspend.c:48. The Backtrace display immediately changes to show the call stack of that thread. So you can make any thread the "active thread" and see its status.

If you set a breakpoint in the data_acq() function, the program will stop when any of the four threads executing data_acq() reaches that point. You can then examine the local variables for that thread. Note, however, that only the local variables for the currently executing thread are visible.

Moving to Kernel Space

Now it's time to rebuild data_acq.c so it runs under RTAI Pthreads. Copy data_acq.c to data_acq_rt.c. main() turns into init_module(). In addition to creating the Command() thread, we now have to explicitly create the Display() thread. printf() changes to printk(). Create a dummy cleanup_module() function.

The major complication of course is that we can't really do any screen I/O from Kernel Space. Among other things, we don't have access to the curses library from Kernel Space. So the real functionality of Display() and Command() needs to be split between Kernel Space and User Space. We can use RTAI FIFOs to communicate between the User Space threads and their "buddies" running in Kernel Space. Remove the curses initialization calls from main()/init_module() and replace with two rtf_create() calls. data_acq.h contains #defines for the two FIFOs.

Kernel Space Display() simply copies the data points it receives to a FIFO. The getstr() call in Command() is just replaced by a rtf_get() call.

The function ftime() is not available in Kernel Space so we'll need an alternative for reading time. RTAI Pthreads defines the following function:

> void clock_gettime (int clock_id, struct timespec *current_time);

This function does not seem to be part of the Posix standards. It doesn't show up in any documentation I have access to. struct timespec is defined in time.h and has two long int fields:

> tv_sec – seconds since midnight Jan. 1, 1970
>
> tv_nsec – nanoseconds since tv_sec

This is almost identical to the timeb structure that data_acq.c uses. clock_id is actually an enum whose only valid value is CLOCK_REALTIME. data_acq() has two calls to ftime(). These need to be replaced with calls to clock_gettime().

RTAI Pthreads has its own version of pthread.h called rtai_pthread.h. Oddly enough, it is not in /usr/src/rtai/include where one would expect, but rather it is in /usr/src/rtai/posix/include. You have a few choices:

- Specify the whole path in the #include directive. Generally not a good idea.

- Add /usr/src/rtai/posix/include to the include path in your compiler flags

- Copy rtai_pthread.h to /usr/src/rtai/include. That's what I did.

logger.c implements the User Space "buddies" for Command() and Display(). Basically it shows what functionality has to be moved out of data_acq_rt.c.

Now open up Makefile. The section that is commented out with '#' builds the kernel module data_acq_rt.o. Delete the '#' characters and run make again.

data_acq_rt.o requires the following modules:

> rtai
>
> rtai_sched
>
> rtai_utils   Provides utility functions for pthread
>
> rtai_pthread
>
> rtai_fifos

The easiest way to get all of these modules loaded is to execute:

> modprobe rtai_pthread
>
> insmod rtai_fifos
>
> insmod ./data_acq_rt.o

rtai_pthread and rtai_fifos are mutually independent so one of them can be loaded with insmod after the other is loaded with modprobe. OK, the second one could be loaded with modprobe too.

Then run ./logger. If everything is working correctly you should be able to enter "c 0 s 500" and see channel 0 start "acquiring data." If not, then it's time to add rt_printk() statements at strategic points in data_acq_rt.c to see what's going on.

Message Queues

Message queues are an element of the overall POSIX standard defined as part of the optional Real-time extensions. The Real-time extensions are not part of standard Linux, at least not in the version 2.4 kernels, but Posix message queues have been implemented in RTAI. So it seems reasonable to include them in this chapter as an element of RTAI's "POSIXness." The corresponding header file is called rtai_pqueue.h and like rtai_pthread.h it is located in /usr/src/rtai/posix/include rather than /usr/src/rtai/include.

Message queues are very similar to RTAI FIFOs except that they are full duplex and they deal in discrete messages rather than continuous streams of bytes. Unfortunately, they only work between tasks in Kernel Space.

```
mqd_t mq_open (char *mq_name, int oflags, mode_t permissions, struct  mq_attr
        *mq_attr);
int mq_close (mqd_t mq);
int mq_unlink (char *mq_name);

size_t mq_receive (mqd_t mq, char *msg_buffer, size_t buflen, unsigned int *msgprio);
int mq_send (mqd_t mq, const char *msg, size_t msglen, unsigned int msgprio);
int mq_notify (mqd_t mq, const struct sigevent *notification);

int mq_getattr (mqd_t mq, struct mq_attr *attrbuf);
int mq_setattr (mqd_t mq, const struct mq_attr *new_attrs, struct mq_attr
        *old_attrs);
```

The message queue API is relatively simple and straightforward. A message queue is opened and closed much like a file. When opening a message queue, you specify access mode and permissions. The open function also has a Pthreads flavor by including an attributes structure. Closing a message queue does not destroy it but merely destroys the link to it. The queue and any messages in it remain available to other links and other threads that may open links to it.

A queue is destroyed, and its resources freed, by *unlinking* it. However, if there are any open links to the queue when mq_unlink() is called, the queue is marked for destruction when the last link is closed. Once a queue has been marked for destruction no further links may be opened to it.

The operations on a queue are just *receive* and *send*. Both operations are blocking unless the O_NONBLOCK flag is set when the link is opened. Messages are placed in the queue in priority order. That is, a higher priority message will be inserted ahead of a lower priority message already there. Messages of equal priority are queued in FIFO order.

mq_notify() allows a thread to specify an asynchronous callback to be called when a message arrives in the queue. The struct sigevent contains, among other things, a pointer to the function to be called. RTAI implements mq_notify() for completeness but does not in fact do asynchronous notification.

The message queue API takes a different approach to attributes than Pthreads. Rather than having a pair of get and set functions for each attribute, the attribute structure fields are made explicitly visible as shown in Listing 11-1. The pair of get and set attributes functions takes a pointer to this structure. When creating a queue you can specify the maximum size of a message and the maximum number of messages that the queue hold. Subsequently the only thing that can be changed by mq_set_attr() is the blocking behavior specified in mq_flags.

Listing 11-1: Message Queue Attribute Structure

```
In file rtai_pqueue.h

//Posix Queue Attributes
struct mq_attr {
    long mq_maxmsg;         //Maximum number of messages in queue
    long mq_msgsize;        //Maximum size of a message (in bytes)
    long mq_flags;        //Blocking/Non-blocking behaviour specifier
                          // not used in mq_open only relevant for
                          // mq_getattrs and mq_setattrs
    long mq_curmsgs;        //Number of messages currently in queue
};
```

The RTAI message queue implementation is another good example of Pthreads programming. Have a look at /usr/src/rtai/posix/src/rtai_pqueue.c.

Suggestions for Further Exploration

If you've managed to get data_acq_rt working and had a look at pqueue.c you've gained a pretty good understanding of the Posix features of RTAI. Consider hacking RTAI itself. Pthreads in RTAI 24.1.9 leaves out a number of useful features. For example:

- What would it take to implement the protocol attribute for mutexes? The basic issue here is changing a task's priority. RTAI doesn't provide an API for this so you would have to get into the guts of rtai_sched.c to understand how to do it.

- Implement the "join" functionality for threads. Here you need to understand what happens when a thread function returns or executes pthread_exit( ).

- It would really be nice if message queues could be used transparently between Kernel Space and User Space like RTAI FIFOs. /usr/src/ rtai/fifos/rtai_fifos.c provides a good starting point. The files in rtai/ lxrt/ offer some ideas about handling queue names, particularly names.c.

Of course, before embarking on any serious update to the RTAI code, you should check the current status. Someone may have already started working on the same project. There's a very active mail list for RTAI. You can subscribe by going to www.rtai.org.

Resources

The Open Group has made available free for private use the entire Posix (now called Single Unix) specification. Go to:

www.unix.org/online.html

You'll be asked to register. Don't worry, it's painless and free. Once you register you can read the specification online or download the entire set of html files for local access.

This chapter has been an introduction to Posix threads. An excellent, more thorough treatment is found in Butenhof, David R., *Programming with POSIX Threads*, Addison-Wesley, 1997.

RTAI Application Programming Interface (API)

This is a summary of the RTAI API as derived from /usr/src/rtai/Documentation/doc_rtai. It is organized by module.

Module: rtai

#include "rtai.h"

Interrupt Management

void rt_global_cli (void);

void rt_global_sti (void);

> rt_global_cli hard disables interrupts (cli) on the requesting cpu and acquires the global spinlock to the calling cpu so that any other cpu synchronized by this method is blocked.

> rt_global_sti hard enables interrupts (sti) on the calling cpu and releases the global lock.

void rt_global_save_flags (unsigned long *flags);

int rt_global_save_flags_and_cli (void);

void rt_global_restore_flags (unsigned long flags);

> rt_global_save_flags saves the cpu interrupt flag (IF) and the global lock flag, in bits 9 and 0 of flags.

rt_global_save_flags_and_cli hard disables interrupts on the requesting CPU and returns old state of CPU interrupt flag (IF) and the global lock flag, in bits 9 and 0.

rt_global_restore_flags restores the cpu hard interrupt flag (IF) and the global lock flag as given by *flags*, freeing or acquiring the global lock according to the state of the global flag bit

void send_ipi_shorthand (unsigned int *shorthand*, int *irq*);

void send_ipi_logical (unsigned long *dest*, int *irq*);

send_ipi_shorthand sends an interprocessor message of *irq* to

- all CPUs if *shorthand* is equal to APIC_DEST_ALLINC;
- all but itself if *shorthand* is equal to APIC_DEST_ALLBUT;
- itself if *shorthand* is equal to APIC_DEST_SELF.

send_ipi_logical sends an interprocessor message of *irq* to all CPUs defined by *dest*. *dest* is given by an unsigned long corresponding to a bitmask of the CPUs to be sent. It is used for local APICs programmed in flat logical mode, so the max number of allowed CPUs is 8, a constraint that is valid for all functions and data of RTAI. The flat logical mode is set when RTAI is installed by calling rt_mount_rtai.

int rt_assign_irq_to_cpu (int *irq*, int *cpu*);

int rt_reset_irq_to_sym_mode (int *irq*);

rt_assign_irq_to_cpu forces the assignment of the external interrupt *irq* to the CPU *cpu*.

rt_reset_irq_to_sym_mode resets the interrupt *irq* to the symmetric interrupts management. The symmetric mode distributes the IRQs over all the CPUs.

Note: These functions are only relevant to multiprocessor systems.

Return value: If there is one CPU in the system, the return value is 1. If there are two or more CPUs, 0 is returned for success. If *cpu* is refers to a nonexistent CPU, the number of CPUs is returned. Other error conditions return -EINVAL.

int rt_request_global_irq (unsigned int *irq*, void (**handler)(void));

int rt_free_global_irq (unsigned int *irq*);

int request_RTirq (unsigned int *irq*, void (**handler)(void));

int free_RTirq (unsigned int *irq*);

> rt_request_global_irq installs function *handler* as an interrupt service routine for IRQ level *irq*. *handler* is then invoked whenever interrupt number *irq* occurs. The installed handler must take care of properly activating any Linux handler using the same *irq* number by calling rt_pend_linux_irq.

> rt_free_global_irq uninstalls the interrupt service routine.

> request_RTirq and free_RTirq are macros defined in rtai.h and is supported only for backwards compatibility with the RTAI variant of RT_linux version 2.0.35. They are fully equivalent to the other two functions above.

Return value:

- 0 – Success
- -EINVAL – *irq* is not a valid IRQ number or *handler* is NULL
- -EBUSY – There is already a handler for interrupt *irq*

int rt_request_linux_irq (unsigned int *irq*, void (**handler)(int *irq*, void **dev_id*, struct pt_regs **regs*), char **linux_handler_id*, void **dev_id*);

int rt_free_linux_irq (unsigned int *irq*, void **dev_id*);

> rt_request_linux_irq installs function *handler* as an interrupt service routine for IRQ level *irq* forcing Linux to share the IRQ with other interrupt handlers. The handler is appended to any already existing Linux handler for the same IRQ and run as a Linux IRQ handler. In this way a real time application can

monitor Linux interrupt handling at is will. The handler appears in

/proc/interrupts

linux_handler_id is a name for /proc/interrupts. The parameter *dev_id* is to pass to the interrupt handler, in the same way as the standard Linux IRQ request call.

The interrupt service routine can be uninstalled with rt_free_linux_irq.

Return value:

- 0 – Success
- -EINVAL – *irq* is not a valid IRQ number or *handler* is NULL
- -EBUSY – There is already a handler for interrupt *irq*

void rt_pend_linux_irq (unsigned int *irq*);

Appends a Linux interrupt *irq* for processing in Linux IRQ mode, i.e. with interrupts fully enabled..

int rt_request_srq (unsigned int *label*, void (**rtai_handler*)(void), long long (**user_handler*)(unsigned int *whatever*));

int rt_free_srq (unsigned int *srq*);

int rt_request_srq. No description provided.

rt_free_srq uninstalls the system call identified by *srq*.

Return value:

- 0 – Success
- -EINVAL – *rtai_handler* is NULL or *srq* is invalid
- -EBUSY – No free srq slot is available.

void rt_pend_linux_srq (unsigned int *srq*);

Appends a system call request *srq* to be used as a service request to the Linux kernel. *srq* is the value returned by rt_request_srq.

Timer Management

void rt_request_timer (void (*handler*)(void), int *tick*, int *apic*);

void rt_free_timer (void);

> rt_request_timer requests a timer of period *tick* ticks, and installs the routine *handler* as a real time interrupt service routine for the timer. Set *tick* to 0 for oneshot mode. (???) If *apic* has a non-zero value the local APIC timer is used. Otherwise timing is based on the 8254.

> rt_free_timer uninstalls the timer previously set by rt_request.

Initialization

void rt_mount_rtai (void);

void rt_umount_rtai (void);

> rt_mount_rtai initializes the real time application interface, i.e. grabs anything related to the hardware, data or service, pointed by at by the Real Time Hardware Abstraction Layer (struct rt_hal rthal;).

> rt_umount_rtai unmounts the real time application interface resetting Linux to its normal state.

Diagnostics

int rt_printk (const char *format*, ...);

> RTAI-safe version of printk().

> **Return value:**
> - > 0 – Number of characters printed

Module: rtai_sched

rtai_sched is a logical link to one of the three possible schedulers, up_sched, smp_sched or mups_sched. Except where noted, all three schedulers support the same API.

#include rtai_sched.h

Task Functions

int rt_task_init (RT_TASK *task*, void (**rt_thread*)(int), int *data*, int
   stack_size, int *priority*, int *uses_fpu*, void(**signal*)(void));

int rt_task_init_cpuid (RT_TASK *task*, void (**rt_thread*)(int), int *data*, int
   stack_size, int *priority*, int *uses_fpu*, void(**signal*)(void), unsigned int
   cpuid);

> Create a real-time task. *task* is a pointer to an RT_TASK type structure whose space must be provided by the application. It must be kept during the whole lifetime of the real-time task and cannot be an automatic variable.
>
> *rt_thread* is the entry point of the task function. The parent task can pass a single integer value *data* to the new task. *stack_size* is the size of the stack to be used by the new task, and *priority* is the priority to be given the task. The highest priority is 0, while the lowest is RT_LOWEST_PRIORITY. *uses_fpu* is a flag. Nonzero value indicates that the task will use the floating point unit.
> *signal* is a function that is called, within the task environment and with interrupts disabled, when the task becomes the current running task after a context switch.
>
> The newly created real-time task is initially in a suspend state. It is can be made active either with rt_task_make_periodic, rt_task_make_periodic_relative_ns or rt_task_resume.
>
> On multiprocessor systems rt_task_init_cpuid assigns task to a specific CPU *cpuid*. rt_task_init automatically selects which CPU the task will run on. This assignment may be changed by calling rt_set_runnable_on_cpus or rt_set_runnable_on_cpuid. If *cpuid* is invalid rt_task_init_cpuid falls back to automatic CPU selection.

Return Value

- 0 – Success

- -EINVAL – Task structure pointed by *task* is already in use.

- -ENOMEM – *stack_size* bytes could not be allocated for the stack.

int rt_task_delete (RT_TASK *task*);

Deletes a real-time task previously created by rt_task_init or rt_task_init_cpuid.

task is the pointer to the task structure.

If task *task* was waiting for a semaphore it is removed from the semaphore waiting queue else any other task blocked on message exchange with *task* is unblocked.

Return Value

- 0 – Success

- -EINVAL – *task* does not refer to a valid task.

int rt_task_make_periodic (RT_TASK *task*, RTIME *start_time*, RTIME *period*);

int rt_task_make_periodic_relative_ns (RT_TASK *task*, RTIME *start_delay*, RTIME *period*);

Mark the task *task*, previously created with rt_task_init, as suitable for a periodic execution, with period *period*, when rt_task_wait_period is called.

The time of first execution is given by *start_time* or *start_delay*. *start_time* is an absolute value measured in clock ticks. *start_delay* is relative to the current time and measured in nanosecs.

Return Value

- 0 – Success

- -EINVAL – *task* does not refer to a valid task.

void rt_task_wait_period (void);

> Suspends the execution of the currently running real-time task until the next period is reached. The task must have been previously marked for execution with rt_task_make_periodic or rt_task_make_periodic_relative_ns.

void rt_task_yield (void);

> Blocks the current task and puts it at the end of the list of ready tasks with the same priority. The scheduler makes the next ready task of the same priority active.

int rt_task_suspend (RT_TASK *task);

> Suspends execution of the task *task*. It will not be executed until a call to rt_task_resume or rt_task_make_periodic is made.

> **Return Value**
> - 0 – Success
> - -EINVAL – *task* does not refer to a valid task.

int rt_task_resume (RT_TASK *task);

> Resumes execution of the task *task* previously suspended by rt_task_suspend or makes a newly created task ready to run.

> **Return Value**
> - 0 – Success
> - -EINVAL – *task* does not refer to a valid task.

int rt_get_task_state (RT_TASK *task);

> Returns the state of a real time task. *task* is a pointer to the task structure.

Return Value: Task state is formed by the bitwise OR of one or more of the following flags:

READY Task *task* is ready to run (i.e. unblocked).

SUSPENDED Task *task* is suspended.

DELAYED Task *task* is waiting for its next running period or expiration of a timeout.

SEMAPHORE Task *task* is blocked on a semaphore.

SEND Task *task* sent a message and is waiting for the receiver task.

RECEIVE Task *task* is waiting for an incoming message.

RPC Task *task* sent a Remote Procedure Call and the receiver has not gotten it yet.

RETURN Task *task* is waiting for reply to a Remote Procedure Call.

Note: the returned task state is only approximate information. Timer and other hardware interrupts may cause a change in the state of the queried task before the caller can evaluate the returned value. Caller should disable interrupts if it wants reliable info about another task.

RT_TASK *rt_whoami (void);

Calling rt_whoami allows a task to get a pointer to its own task structure.

Return value: Pointer to the currently running task.

int rt_task_signal_handler (RT_TASK *task*, void (*handler*)(void));

Installs or changes the signal function of a real-time task.
task is a pointer to the real time-task
handler is the entry point of the signal function.

Signal handler is a function called within the task environment and with interrupts disabled, when the task becomes the current running task after a context switch. The signal handler function can also be set when the task is created with rt_task_init.

Return Value
- 0 – Success
- -EINVAL – *task* does not refer to a valid task.

void rt_set_runnable_on_cpus (RT_TASK *task*, unsigned int *cpu_mask*);

void rt_set_runnable_on_cpuid (RT_TASK *task*, unsigned int *cpuid*);

Select one or more CPUs that are allowed to run task *task*. rt_set_runnable_on_cpuid assigns task to a specific CPU while rt_set_runnable_on_cpus magically selects one CPU from the given set that task *task* will run on. Bit<n> of *cpu_mask* enables CPU<n>.

If no CPU selected by *cpu_mask* or *cpuid* is available, both functions automatically select a possible CPU.

Note: This call has no effect on uniprocessor systems.

int rt_task_use_fpu (RT_TASK* *task*, int *use_fpu_flag*);

void rt_linux_use_fpu (int *use_fpu_flag*);

rt_task_use_fpu informs the scheduler that the real time task *task* will use floating point arithmetic operations.

rt_linux_use_fpu informs the scheduler that user space Linux processes will use floating point arithmetic operations.

If *use_fpu_flag* has nonzero value, FPU context is also switched when *task* or the kernel becomes active. This makes task switching slower. Initial value of this flag is set by rt_task_init when the real-time task is created. By default Linux "task" has this flag cleared. It can be set with LinuxFpu command line parameter of the rtai_sched module.

Return Value (rt_task_use_fpu)

- 0 – Success
- -EINVAL – *task* does not refer to a valid task.

void rt_preempt_always (int *yes_no*);

void rt_preempt_always_cpuid (int *yes_no*, unsigned int*cpu_id*);

In the one-shot mode a timed task is made active/current at the expiration of the timer shot. The next timer expiration is programmed by choosing among the timed tasks the one with a priority higher than the current after the current has released the CPU, always assuring the Linux timing. While this policy minimizes the programming of the one-shot mode, enhancing efficiency, it can be unsuitable when a task has to be guarded against looping by watch dog task with high-priority value, as in such a case the latter has no chance of running.

Calling these functions with nonzero value assures that a timed high-priority preempting task is always programmed to be fired while another task is current. The default is no immediate preemption in one-shot mode, firing of the next shot programmed only after the current task releases the CPU.

Initial value of this flag can be set with PreemptAlways command line parameter of the rtai_sched module.

Note: currently both functions are identical. Parameter *cpu_id* is ignored.

Timer Functions

void rt_set_oneshot_mode (void);

void rt_set_periodic_mode (void);

> rt_set_oneshot_mode sets the timer to one-shot timing mode. It consists in a variable timing based on the cpu clock frequency. This allows task to be timed arbitrarily. It must be called before using any time related function, including conversions.

> rt_set_periodic_mode sets the timer to periodic timing mode. It consists of a fixed frequency timing of the tasks in multiple of the period set with a call to start_rt_timer. The resolution is that of the 8254 frequency (1193180 hz). Any timing request not an integer multiple of the period is satisfied at the closest period tick. This is the default mode.

> One-shot mode can also be set initially with the OneShot command line parameter of the rtai_sched module

> Note: Stopping the timer by stop_rt_timer sets the timer back into its default (periodic) mode. Call rt_set_oneshot_mode before each start_rt_timer if it required.

RTIME start_rt_timer (int *period*);

void stop_rt_timer (void);

> start_rt_timer starts the timer with a period *period*. The period is in internal count units and is required only for the periodic mode. In the one-shot the parameter value is ignored.

> stop_rt_timer stops the timer. The timer mode is set to periodic

RTIME count2nano (RTIME *timercounts*);

RTIME nano2count (RTIME *nanosecs*);

> count2nano converts the time of *timercounts* internal count units into nanoseconds.

nano2count converts the time of *nanosecs* nanoseconds into internal counts units.

Remember that the count units are related to the cpu frequency in one-shot mode and to the 8254 frequency (1193180 Hz) in periodic mode.

RTIME rt_get_time (void);

RTIME rt_get_time_ns (void);

RTIME rt_get_cpu_time_ns (void);

> rt_get_time returns the number of real time clock ticks since RT_TIMER bootup (*whatever this means*). This number is multiple of the 8254 period in periodic mode, while is multiple of cpu clock period in one-shot mode.
>
> rt_get_time_ns is the same as rt_get_time but the returned time is converted to nanoseconds.
>
> rt_get_cpu_time_ns ???

RTIME next_period (void);

> Returns the time when the caller task will run next. This is only relevant for periodic tasks.

void rt_busy_sleep (int *nanosecs*);

void rt_sleep (RTIME *delay*);

void rt_sleep_until (RTIME *time*);

> rt_busy_sleep delays the execution of the caller task without giving back the control to the scheduler. This function burns up CPU cycles in a busy wait loop. It should be used for very short synchronization delays only.
> *nanosecs* is the number of nanoseconds to wait.
>
> rt_sleep suspends execution of the caller task for a time of *delay* internal count units. During this time the CPU is used by other tasks.

rt_sleep_until is similar to rt_sleep but the parameter *time* is the absolute time when the task should wake up. If the given time is already passed this call has no effect

Note: a higher priority task or interrupt handler can run during wait so the actual time spent in these functions may be longer than that specified.

Semaphore Functions

All of the communication and synchronization mechanisms in rtai_sched have a similar API.

void rt_sem_init (SEM* *sem*, int *value*);

Initializes a semaphore *sem*. A semaphore can be used for communication and synchronization among real-time tasks. *sem* must point to a statically allocated structure. *value* is the initial value of the semaphore (usually 1). The initial value must be non-negative.

A positive value of the semaphore variable shows how many tasks can do a wait operation without blocking. (Say how many tasks can enter the critical region.) A negative semaphore value shows how many tasks are blocked on it.

int rt_sem_delete (SEM* *sem*);

Deletes a semaphore previously created with rt_sem_create. *sem* points to the structure used in the corresponding call to rt_sem_create.

Any tasks blocked on this semaphore are allowed to run when the semaphore is deleted

Return Value

- 0 – Success

- SEM_ERR – *sem* does not refer to a valid semaphore. (-EINVAL would be a more consistent return value)

int rt_sem_signal (SEM* *sem*);

This is the semaphore post (sometimes known as 'give', 'signal', or 'V') operation. It is typically called when the task leaves a critical region. The semaphore value is incremented and tested. If the value is not positive, the first task in semaphore's waiting queue is allowed to run. rt_sem_signal does not block the caller task.

Return Value

- 0 – Success

- SEM_ERR – *sem* does not refer to a valid semaphore.

int rt_sem_wait (SEM* *sem*);

This is the semaphore pend (sometimes known as 'take', 'wait' or 'P') operation. It is typically called when a task enters a critical region. The semaphore value is decremented and tested. If it is still non-negative rt_sem_wait returns immediately. Otherwise the caller task is blocked and queued up. Queueing may happen in either priority order or FIFO order as determined by the compile time option SEM_PRIORD. In this case rt_sem_wait returns if

- The caller task is at the head of the waiting queue and another task issues a rt_sem_signal;

- An error occurs (e.g. the semaphore is destroyed);

Return Value

- On success an indeterminate value that somehow depends on the value of the semaphore is returned. This should be considered a bug.

- SEM_ERR – *sem* does not refer to a valid semaphore.

int rt_sem_wait_if (SEM* *sem*);

A version of rt_sem_wait that never blocks the caller. The return value indicates whether or not the calling task "got" the semaphore.

Return Value

- 0 – Semaphore was not available.

- > 0 – "Previous" value of the semaphore. The semaphore has been decremented.

- SEM_ERR – *sem* does not refer to a valid semaphore.

int rt_sem_wait_until (SEM* *sem*, RTIME *time*);

int rt_sem_wait_timed (SEM* *sem*, RTIME *delay*);

These are timed versions of rt_sem_wait. If the current semaphore value is less than 0 and the specified time interval expires before another task posts to the semaphore, these calls return with an error. rt_sem_wait_timed waits for up to *delay* internal counts. rt_sem_wait_until waits until an absolute time.

Return Value

- On success an indeterminate value that somehow depends on the value of the semaphore is returned. This should be considered a bug.

- SEM_ERR – *sem* does not refer to a valid semaphore.

- SEM_TIMEOUT – The specified interval expired before the semaphore became available.

Mailbox functions

Mailboxes are a flexible method of task-to-task communication. Tasks are allowed to send arbitrary size messages by using any mailbox buffer size. Clearly you should use a buffer sized at least as big as the largest message you envision. However, if you expect a message larger than the average message size very rarely, you can use a smaller buffer without much loss of efficiency. In such a way you can set up your own mailbox usage protocol, e.g. using fixed size messages with a buffer that is an integer multiple of such a size guarantees that each message is sent/received automatically to/from the mailbox. Multiple senders and receivers are allowed and each will get the service it requires in turn, according to its priority.

int rt_mbx_init (MBX* *mbx*, int *size*);

> Initializes a mailbox of size *size*. *mbx* points to a statically allocated mailbox structure. RTAI will dynamically allocate the buffer space.

Return Value

- 0 – Success
- -EINVAL – Space could not be allocated for the mailbox buffer.

int rt_mbx_delete (MBX* *mbx*);

> Removes a mailbox previously created with rt_mbox_init. *mbx* points to the structure used in the corresponding call to rt_mbox_init.

Return Value

- 0 – Success
- -EINVAL – *mbx* does not point to a valid mailbox.

int rt_mbx_send (MBX* *mbx*, void* *msg*, int *msg_size*);

> Sends a message *msg* of *msg_size* bytes to the mailbox *mbx*. The caller will be blocked until the whole message is enqueued or an error occurs.

> ### Return Value
>
> - 0 – Success
> - -EINVAL – *mbx* does not point to a valid mailbox.

int rt_mbx_send_wp (MBX* *mbx*, void* *msg*, int *msg_size*);

int rt_mbx_send_if (MBX* *mbx*, void* *msg*, int *msg_size*);

> Non-blocking versions of rt_mbx_send.

> rt_mbx_send_wp sends as much as possible of message *msg* to mailbox *mbx* then returns immediately. "wp" means send "what's possible".

> rt_mbx_send_if sends message *msg* to the mailbox *mbx* only if the entire message can be enqueued without blocking. Otherwise it returns an error.

> ### Return Value
>
> - >= 0 – Success. A non-zero value is the number of bytes of the message *not* sent.
> - -EINVAL – *mbx* does not point to a valid mailbox.

int rt_mbx_send_until (MBX* *mbx*, void* *msg*, int *msg_size*, RTIME *time*);

int rt_mbx_send_timed (MBX* *mbx*, void* *msg*, int *msg_size*, RTIME *delay*);

> Timed versions of rt_mbx_send. These functions return after the specified time interval expires whether or not the entire message has been sent.

> rt_mbx_send_until. waits until an absolute time.

> rt_mbx_send_timed waits for up to *delay* internal counts.

Return Value

- >= 0 – A nonzero value is the number of bytes of the message *not* sent.

- -EINVAL – *mbx* does not point to a valid mailbox.

int rt_mbx_receive (MBX* *mbx*, void* *msg*, int *msg_size*);

Receives a message from the mailbox *mbx*. *msg* points to a buffer of *msg_size* bytes provided by the caller. The caller will be blocked until all bytes of the message arrive or an error occurs.

Return Value

- >= 0 – Number of bytes received from mailbox.

- -EINVAL – *mbx* does not point to a valid mailbox.

int rt_mbx_receive_wp (MBX* *mbx*, void* *msg*, int *msg_size*);
int rt_mbx_receive_if (MBX* *mbx*, void* *msg*, int *msg_size*),

Non-blocking versions of rt_mbx_receive.

rt_mbx_receive_wp receives at most *msg_size* bytes of message from mailbox *mbx* then returns immediately.

rt_mbx_receive_if receives a message from the mailbox *mbx* if the whole message of *msg_size* bytes is available immediately.

Return Value

- >= 0 – Number of bytes received from mailbox.

- -EINVAL – *mbx* does not point to a valid mailbox.

int rt_mbx_receive_until (MBX* *mbx*, void* *msg*, int *msg_size*, RTIME *time*);
int rt_mbx_receive_timed (MBX* *mbx*, void* *msg*, int *msg_size*, RTIME *delay*);

Timed versions of rt_mbx_receive. These functions return after the specified time interval expires whether or not the entire message has been sent.

rt_mbx_receive_until. waits until an absolute time.

rt_mbx_receive_timed waits for up to *delay* internal counts.

Return Value

- \>= 0 – Number of bytes received from mailbox.
- -EINVAL – *mbx* does not point to a valid mailbox.

Message Handling Functions

This set of functions implements a direct task-to-task messaging mechanism. The message content is restricted to a single integer.

RT_TASK* rt_send (RT_TASK* *task*, unsigned int *msg*);

RT_TASK* rt_send_if (RT_TASK* *task*, unsigned int *msg*);

> Sends the message *msg* to the task *task*. rt_send blocks the calling task until the destination task, *task*, gets the message. If multiple tasks are sending messages to *task*, they are queued in either priority order or FIFO order as determined by compile time option MSG_PRIORD. rt_send_if doesn't block.

Return value

- Success – *task* (the pointer to the task that received the message)
- 0 –   rt_send: *task* was killed before it could receive the message.

  rt_send_if: *task* was not ready to receive the message
- MSG_ERR – *task* does not reference a valid task.

RT_TASK* rt_send_until (RT_TASK* *task*, unsigned int *msg*, RTIME *time*);

RT_TASK* rt_send_timed (RT_TASK* *task*, unsigned int *msg*, RTIME *delay*);

Timed versions of rt_ send. These functions return after the specified time interval expires whether or not *msg* was successfully sent.

Return value

- Success – *task* (the pointer to the task that received the message)
- 0 – Operation timed out. Message was not delivered.
- MSG_ERR – *task* does not reference a valid task.

RT_TASK* rt_receive (RT_TASK* *task*, unsigned int **msg*);

RT_TASK* rt_receive_if (RT_TASK* *task*, unsigned int **msg*);

Gets a message from the task specified by *task*. If *task* is equal to 0, the caller accepts a message from any task. If there is a pending message, rt_receive returns immediately. Otherwise the caller task is blocked and queued up. (Queueing may happen in priority order or FIFO order as determined by compile time option MSG_PRIORD.) *msg* points to a buffer provided by the caller. rt_receive_if returns immediately whether or not a message is pending.

Return value

- Success – *task* (the pointer to the task that sent the message)
- 0 –   rt_receive: *task* was killed before it could send the message.

    rt_receive_if: no message was sent.
- MSG_ERR – *task* does not reference a valid task.

RT_TASK* rt_receive_until (RT_TASK* *task*, unsigned int **msg*, RTIME *time*);

RT_TASK* rt_receive_timed (RT_TASK* *task*, unsigned int **msg*, RTIME *delay*);

Timed versions of rt_ receive. These functions return after the specified time interval expires whether or not *msg* was successfully received.

Return value

- Success – *task* (the pointer to the task that received the message)
- 0 – Operation timed out. Message was not received.
- MSG_ERR – *task* does not reference a valid task.

Remote Procedure Calls

Although this mechanism is called "Remote Procedure Calls," in reality it is just a full duplex version of the message handling described above.

RT_TASK *rt_rpc (RT_TASK *task*, unsigned int *msg*, unsigned int **reply*);

RT_TASK *rt_rpc_if (RT_TASK *task*, unsigned int *msg*, unsigned int **reply*);

Makes a Remote Procedure Call. RPC is like a send/receive pair. rt_rpc sends the message *msg* to the task *task* then waits until a reply is received. The caller task is always blocked and queued up. (Queueing may happen in priority order or FIFO order as determined by compile time option MSG_PRIORD.) The receiver task may get the message with any rt_receive_* function. It can send the answer with rt_return. *reply* points to a buffer provided by the caller. rt_rpc_if doesn't block.

Return value

- Success – *task* (the pointer to the task that received the message)
- 0 – *task* was not ready to receive the message (rt_rpc_if) or receiver task was killed before receiving the message.
- MSG_ERR – *task* does not reference a valid task.

RT_TASK *rt_rpc_until (RT_TASK *task*, unsigned int *msg*, unsigned int *reply*, RTIME *time*);

RT_TASK *rt_rpc_timed (RT_TASK *task*, unsigned int *msg*, unsigned int *reply*, RTIME *delay*);

> Timed versions of rt_ rpc. These functions return after the specified time interval expires whether or not *msg* was successfully sent.

> **Return value**
> - Success – *task* (the pointer to the task that received the message)
> - 0 – Operation timed out. Message was not delivered.
> - MSG_ERR – *task* does not reference a valid task.

RT TASK *rt_return (RT_TASK *task*, unsigned int *result*);

> Sends *result* back to *task*. If the task calling **rt_rpc_*** previously is not waiting for the answer (i.e. killed or timed out) this return message is silently discarded.

> **Return value**
> - Success – *task* (the pointer to the task that received the reply)
> - 0 – Reply was not delivered.
> - MSG_ERR – *task* does not reference a valid task.

int rt_isrpc (RT_TASK *task*);

> After receiving a message, a task can determine whether the sender *task* is waiting for a reply or not by calling rt_isrpc. No answer is required if the message sent by a rt_send_* function or the sender called rt_rpc_timed or rt_rpc_until but it is already timed out.

Return value

- 0 – *msg* was sent by rt_send_*, no reply necessary.
- Non-zero – *task* is expecting a reply.

rt_isrpc is not necessary because rt_return is smart enough to determine if a reply is required. Use of rt_isrpc is discouraged.

Module rtai_fifos

#include <rtai_fifos.h>

rtai_fifos provides a point-to-point sequenced mechanism for communicating between Kernel Space real-time tasks and User Space processes. The Kernel Space API is described here. User Space processes treat RTAI FIFOs as character devices /dev/rtf*n*. These devices are accessed from User Space using the normal open( ), read( ) and write( ) system calls.

Reading and writing to fifos in Kernel Space is non-blocking.

int rtf_create (unsigned int *fifo*, int *size*);

> Creates a real-time fifo (RT-FIFO) of initial size *size* and assigns it the identifier *fifo*.
> *fifo* is a small positive integer that identifies the fifo on further operations. It must be less than RTF_NO. *fifo* may refer an existing RT-FIFO. In this case the size is adjusted if necessary.

Return value

- Success – *size* (the argument passed to rtf_create)
- -ENODEV – *fifo* is greater than or equal to RTF_NO.
- -ENOMEM – *size* bytes could not be allocated for the RT-FIFO.

int rtf_destroy (unsigned int *fifo*);

>Closes a real-time *fifo* previously created/reopened with rtf_create( ) or rtf_open_sized( ). An internal mechanism counts how many times a *fifo* was opened. Opens and closes must be in pair. rtf_destroy( ) should be called as many times as rtf_create( ) was. After the last close the *fifo* is really destroyed.

>**Return value**

>- Success – The number of real-time tasks still having this FIFO open. Zero means the FIFO really was closed.
>- -ENODEV – *fifo* is greater than or equal to RTF_NO.
>- -EINVAL – *fifo* is not a valid open FIFO.

int rtf_reset (unsigned int *fifo*);

>Removes any data posted to, but not yet removed from, *fifo*. The successful result is that *fifo* contains no data.

>**Return value**

>- Success – 0.
>- -ENODEV – *fifo* is greater than or equal to RTF_NO.
>- -EINVAL – *fifo* is not a valid open FIFO.

int rtf_resize (unsigned int *fifo*, int *size*);

>Modifies the real-time fifo *fifo*, previously created with rtf_create( ), to have a new size of *size*. Any data currently in *fifo* is discarded.

>**Return value**

>- Success – *size* (the argument passed to rtf_resize)
>- -ENODEV – *fifo* is greater than or equal to RTF_NO.
>- -EINVAL – *fifo* is not a valid open FIFO.

- -ENOMEM – *size* bytes could not be allocated for the RT-FIFO.

int rtf_put (unsigned int *fifo*, void **buf*, int *count*);

> Write a block of data to a real-time fifo previously created with rtf_create. *fifo* is the ID with which the RT-FIFO was created. *buf* points the block of data to be written. *count* is the size of the block in bytes. This mechanism is available only to real-time tasks; Linux processes use a write() to the corresponding /dev/fifo<*n*> device to enqueue data to a fifo.

> **Return value**
>
> - Success – the number of bytes written. Note that this value may be less than *count* if *count* bytes of free space is not available in the fifo
> - -ENODEV – *fifo* is greater than or equal to RTF_NO.
> - -EINVAL – *fifo* is not a valid open FIFO.

int rtf_get (unsigned int *fifo*, void **buf*, int *count*);

> Reads a block of data from a real-time fifo previously created with a call to rtf_create. *fifo* is the ID with which the RT-FIFO was created. *buf* points a buffer of *count* bytes size provided by the caller. This mechanism is available only to real-time tasks; Linux processes use a read() from the corresponding fifo device to dequeue data from a fifo.

> **Return value**
>
> - Success – the number of bytes read. Note that this value may be less than *count* if *count* bytes were not available in the fifo at the time of the call
> - -ENODEV – *fifo* is greater than or equal to RTF_NO.
> - -EINVAL – *fifo* is not a valid open FIFO.

int rtf_create_handler (unsigned int *fifo*, int (*handler)(unsigned int *fifo*));

int rtf_create_handler (unsigned int *fifo*, X_FIFO_HANDLER(*handler*));

> Installs a handler that is executed when data is written to or read from a real-time fifo. *fifo* is an RT-FIFO that must have previously been created with a call to rtf_create. The function pointed by *handler* is called whenever a Linux process accesses that fifo. The X_FIFO_HANDLER form allows for an extended handler function prototyped as:
>
> > int (*handler)(unsigned int *fifo*, int *rw*)
>
> This allows the handler to determine whether it was called as the result of a read (rw = 'r') or a write (rw = 'w').
>
> rtf_create_handler is often used in conjunction with rtf_get to process data acquired asynchronously from a Linux process. The installed handler calls rtf_get when data is present. Because the handler is only executed when there is activity on the fifo, polling is not necessary

> **Return value**
> - Success 0
> - -EINVAL – *fifo* is not a valid open FIFO.

Module rtai_shm

#include <rtai_shm.h>

rtai_shm supports memory regions shared between Kernel Space RTAI tasks and User Space processes.

Kernel Space API

void *rtai_kmalloc (unsigned long *name*, int *size*);

> Allocates a shared memory region named *name* with size *size* bytes. The first call to rtai_kmalloc for a given name allocates

the space. Subsequent calls with the same name, whether from Kernel Space or User Space, just attach to it.

Return value

- Success – Address of allocated space
- 0 – Unable to allocate requested space

void rtai_kfree (void *adr*);

Frees the shared memory region identified by *name*. Actually, the region is just unmapped until the last process/task attached to this region calls rtai_kfree. Then the memory space is freed.

User Space API

void *rtai_malloc (unsigned long *name*, int *size*);

void *rtai_malloc_adr (void *adr*, unsigned long *name*, int *size*);

Allocates a shared memory region named *name* with size *size* bytes. The first call to rtai_malloc or rtai_malloc_adr for a given name allocates the space. Subsequent calls with the same name, whether from Kernel Space or User Space, just attach to it.

rtai_malloc lets the system assign the address.

rtai_malloc_adr requests a specific address. This is just a "suggestion." The returned value may or may not be the same as *adr*.

Return value

- Success – Address of allocated space
- 0 – Unable to allocate requested space

void rtai_free (unsigned long *name,* void *adr*);

Frees the shared memory region identified by *name*. Actually, the region is just unmapped until the last process/task attached to this region calls rtai_free. Then the memory space is freed.

Utility Functions

These functions are also used by LXRT.

unsigned long nam2num (const char *name);

> Translates an ASCII *name* of up to six characters to an unsigned long id. The valid character set is:
>
> > 'A' to 'Z', 'a' to 'z' (case not preserved)
> >
> > '0' to '9', '_'
>
> All other characters are translated to a value that is converted back to '$' by num2nam.

> **Return value**
>
> - Success – the converted id.
> - There is no error condition

void num2nam (unsigned long *id*, char *name*);

> Translates *id* back to an ASCII *name* string using the same algorithm as nam2num.

Module rtai_lxrt

#include <rtai_lxrt.h>

Allows the RTAI API to be used (almost) transparently from User Space processes. This section describes only the functions that differ from the Kernel Space API. All other functions are used as described in previous sections.

Object Initialization

RT_TASK *rt_task_init (unsigned int *id*, int *priority*, int *stack_size*, int *max_msgsize*);

> Creates a new real-time task in User Space named *id*, with priority *priority*. *stack_size* and *max_msgsize* may be 0 in which case default values are used. Default stack size is 512 bytes, default max message size is 256 bytes.
>
> **Return value**
>
> - Success – pointer to a task structure in Kernel Space. Note that this value must not be used directly in User Space. It may only be passed as an argument to other LXRT functions.
> - 0 – Unable to create buddy task or *id* is already registered in the name space

SEM *rt_sem_init (unsigned long *id*, int *initial_count*);

> Allocates and initializes a semaphore with name *id* and initial value *initial_count*.
>
> **Return value**
>
> - Success – pointer to a semaphore structure in Kernel Space. Note that this value must not be used directly in User Space. It may only be passed as an argument to other LXRT functions.
> - 0 – Unable to create semaphore or *id* is already registered in the name space

MBX *rt_mbx_init (unsigned long *id*, int *size*);

> Allocates and initializes a mailbox with name *id* and size *size*.

Return value

- Success – pointer to a mailbox structure in Kernel Space. Note that this value must not be used directly in User Space. It may only be passed as an argument to other LXRT functions.

- 0 – Unable to create mailbox or *id* is already registered in the name space.

Kernel Space Namespace Utilities

Functions to manage the LXRT namespace from kernel modules.

Note: These functions are not prototyped in any header file. Their use will generate a compiler warning.

int rt_register (unsigned long *id*, void **adr*);

> Associates the name *id* with the Kernel Space object pointed to by *adr* and registers this object in the LXRT namespace. This allows User Space LXRT processes to reference the object.

Return value

- Success – positive integer. Actually the slot number in the namespace table.

- 0 – Unable to register the object. Namespace table is full

int rt_drg_on_adr (void **adr*);

int rt_drg_on_name (unsigned long *id*);

> Remove an object from the namespace. The object to be removed can be referenced either by address (*adr*) or name (*id*).

Return value

- Success – positive integer. Actually the slot number in the namespace table that the object occupied.

- 0 – Unable to de-register the object. Not found in namespace table.

User Space Namespace Utilities

These functions allow User Space LXRT processes to access Kernel Space objects.

void *rt_get_adr (unsigned long *id*);

Returns the address of the object named *id*.

unsigned long rt_get_name (void *adr*);

Returns the name of the object at address *adr*.

Return value

- Both functions return zero if the requested object is not in the namespace.

User Space Hard Real-time

void rt_make_hard_real_time (void);

Gives a User Space process hard real-time characteristics. The process must be locked in memory. Hard real-time processes should avoid making any Linux system call that could cause a task switch. Note that only processes run by the root user can lock memory (however, see rt_allow_nonroot_hrt() below).

void rt_make_soft_real_time (void);

Returns a process back to normal soft real-time behavior.

void rt_allow_nonroot_hrt (void);

Allows a non-root user to lock memory and invoke hard real-time.

Posix Threads (Pthreads) Application Programming Interface

This is a summary of the features of Posix 1003.1c-1995 supported under RTAI.

Unless indicated otherwise, all Pthreads functions return an integer status code where zero indicates that the function succeeded and a negative number indicates an error condition. Pthreads makes a distinction between two categories of errors:

1. Mandatory ("if occurs") errors involve circumstances beyond the control of the programmer. You wouldn't be expected to know, for example, or even be able to determine, that there isn't sufficient virtual memory to create a new thread. So the system must always detect and report this kind of error.

2. Optional ("If detected") errors are conditions that are usually your mistake. Attempting to lock a mutex that hasn't been initialized or trying to unlock a mutex that is locked by another thread are examples of this category. It may simply be too expensive in terms of processor time or other system resources to detect and report all of these potential error conditions. A competent programmer should be able to detect and solve these problems without help from the system.

Some systems may provide a "debugging" mode that turns on some or all of these optional errors while you're developing code. When production code is ready for release, you turn off the debugging mode.

In the following descriptions, mandatory error conditions are indicated in **bold**. For function arguments that take symbolic values, the list of valid symbols is given and the default value is indicated in **bold**.

> #include <pthread.h>   // for User Space pthreads
>
> #include <rtai_pthread.h>   // for Kernel Space pthreads using RTAI

Threads

int pthread_create (pthread_t *tid*, const pthread_attr_t *attr*, void
    *(*start*) (void *), void *arg*);

> Creates a new thread with optional creation attributes *attr*.
> The new thread executes the function *start* with argument *arg*.
> *tid* is the thread's ID or handle.
>
> **Errors**
>
> - **EINVAL** – *attr* invalid
> - **EAGAIN** – insufficient resources available to create the thread

int pthread_exit (void *ret_val*);

> Terminates the calling thread after first calling any registered cleanup handlers. *ret_val* is returned to any thread joining this one.

pthread_t pthread_self (void);

> Returns the ID of the calling thread.

int sched_yield (void);

> Yields the processor and makes the thread ready only after all other ready threads at this priority level have run.
>
> Note: This function is from Posix 1.b (real-time extensions) and is declared in <sched.h>. It reports errors by setting the error code in errno and returning −1 as its value

Errors

- **ENOSYS** – sched_yield not supported.

Thread Attributes

int pthread_attr_init (pthread_attr_t *attr*);

> Initializes a pthread attributes object with default values.
>
> Errors
>
> - **ENOMEM** – insufficient memory for attribute object

int pthread_attr_destroy (pthread_attr_t *attr*);

> Destroys an attribute object. Note that this does not affect threads previously created using *attr*.
>
> Errors
>
> - EINVAL – *attr* is not a thread attribute object

int pthread_attr_getdetachstate (const pthread_attr_t *attr*, int *detach_state*);

int pthread_attr_setdetachstate (pthread_attr_t *attr*, int *detach_state*);

> Get or set the thread's "detach" state. Detached threads can't be cancelled or joined.

> *detach_state* = **PTHREAD_CREATE_JOINABLE**
>
> > PTHREAD_CREATE_DETACHED
>
> Errors
>
> - EINVAL – *attr* is not a thread attribute object
> - EINVAL – *detach_state* is invalid (set only)

Scheduling Policy Attributes

int pthread_attr_getinheritsched (const pthread_attr_t *attr*, int *inherit_sched*);

int pthread_attr_setinheritsched (pthread_attr_t *attr*, int *inherit_sched*);

Determines whether threads created with this *attr* inherit the scheduling policy and parameters of the calling creator or the parameters in *attr*.

inherit_sched = PTHREAD_INHERIT_SCHED

PTHREAD_EXPLICIT_SCHED

Default value is implementation-dependend.

Errors

- **ENOSYS** – Priority scheduling is not supported
- EINVAL – *attr* is not a thread attribute object or *inherit_sched* invalid (set only)

int pthread_attr_getschedparam (const pthread_attr_t *attr*, struct sched_param *sched_p*);

int pthread_attr_setschedparam (pthread_attr_t *attr*, struct sched_param *sched_p*);

Gets or sets the scheduling parameters used by threads created with this *attr*. Contents of *sched_p* are implementation-dependent.

Errors

- **ENOSYS** – Priority scheduling is not supported
- EINVAL – *attr* is not a thread attribute object or *sched_p* invalid (set only)

int pthread_attr_getschedpolicy (const pthread_attr_t *attr*, int *policy*);

int pthread_attr_setschedpolicy (pthread_attr_t *attr*, int *policy*);

Gets or sets the scheduling policy used by threads created with this *attr*.

Policy = SCHED_FIFO

SCHED_RR

SCHED_OTHER

Default value is implementation-dependent.

Errors

- **ENOSYS** – Priority scheduling is not supported
- **EINVAL** – *attr* is not a thread attribute object or *policy* invalid (set only)

int pthread_attr_getscope (const pthread_attr_t *attr*, int **scope*);

int pthread_attr_setscope (pthread_attr_t *attr*, int *scope*);

Gets or sets the "contention scope" of threads created with this *attr*.

> *scope* =   PTHREAD_SCOPE_PROCESS
>
>   PTHREAD_SCOPE_SYSTEM

Default value is implementation-dependent.

Errors

- **ENOSYS** – Priority scheduling is not supported
- **EINVAL** – *attr* is not a thread attribute object or *scope* invalid (set only)

Mutexes

int pthread_mutex_init (pthread_mutex_t *mutex*, const pthread_mutexattr_t *mutex_attr*);

Creates a *mutex* object. If non-null, *mutex_attr* provides optional attributes.

Errors

- **EAGAIN** – insufficient resources other than memory
- **ENOMEM** – insufficient memory for mutex object
- **EPERM** – no privilege to perform this operation
- EBUSY – *mutex* is already initialized
- EINVAL – *mutex_attr* is not a mutex attribute object

int pthread_mutex_destroy (pthread_mutex_t *mutex*);

> Destroys an existing *mutex* that is no longer needed.
>
> **Errors**
>
> - EBUSY – *mutex* is in use
> - EINVAL – *mutex* is not a mutex

int pthread_mutex_lock (pthread_mutex_t *mutex*);

> Locks a mutex. If *mutex* is already locked, the calling thread is blocked until *mutex* is subsequently unlocked. On return the calling thread "owns" *mutex* until it calls pthread_mutex_unlock().
>
> **Errors**
>
> - **EINVAL** – thread priority exceeds *mutex* priority ceiling.
> - EINVAL – *mutex* is not a mutex object
> - EDEADLK – calling thread already owns *mutex*

int pthread_mutex_trylock (pthread_mutex_t *mutex*);

> Locks *mutex* if it is currently unlocked. If *mutex* is locked return immediately with error code. This is a non-blocking method of locking a mutex.
>
> **Errors**
>
> - **EINVAL** – thread priority exceeds *mutex* priority ceiling.
> - **EBUSY** – *mutex* is already locked
> - EINVAL – *mutex* is not a mutex object
> - EDEADLK – calling thread already owns *mutex*

int pthread_mutex_unlock (pthread_mutex_t *mutex*);

> Unlocks *mutex*. If any threads are waiting on this mutex, one of them is awakened and becomes the new owner. The order in which threads are awakened depends on scheduling policy.

Errors

- EINVAL – *mutex* is not a mutex
- EPERM – calling thread does not own *mutex*

Mutex Attributes

int pthread_mutexattr_init (pthread_mutexattr_t *mutex_attr);

> Initializes a mutex attributes object with default values.
>
> **Errors**
>
> - **ENOMEM** – insufficient memory for attribute object

int pthread_mutexattr_destroy (pthread_mutexattr_t *mutex_attr);

> Destroys a mutex attribute object.
>
> **Errors**
>
> - EINVAL – *mutex_attr* is not a mutex attribute object

int pthread_mutexattr_getkind_np (const pthread_mutexattr_t *mutex_attr, int *kind);

int pthread_mutexattr_setkind_np (pthread_mutexattr_t *mutex_attr, int kind);

> Gets or sets the mutex "kind" or type. This is a non-portable Linux extension. See Chapter 11 for details on mutex kind.
>
> *kind* =   **PTHREAD_MUTEX_FAST_NP**
>   PTHREAD_MUTEX_RECURSIVE_NP
>   PTHREAD_MUTEX_ERRORCHECK_NP
>
> **Errors**
>
> - EINVAL – *mutex_attr* is not a mutex attribute object
> - **EINVAL** – *kind* is invalid (set only)

int pthread_mutexattr_getprioceiling (const pthread_mutexattr_t
  mutex_attr, int *prioceiling*);

int pthread_mutexattr_setprioceiling (pthread_mutexattr_t *mutex_attr*, int
  prioceiling);

> Gets or sets the priority ceiling at which threads run while owning a mutex created with *attr*. Not implemented in RTAI Pthreads.
>
> **Errors**
>
> - **ENOSYS** – Priority scheduling is not supported
> - EINVAL – *attr* is not a mutex attribute object or *prioceiling* invalid (set only)
> - ENOPERM – no permission to set *prioceiling*

int pthread_mutexattr_getprotocol (const pthread_mutexattr_t
  mutex_attr, int *protocol*);

int pthread_mutexattr_setprotocol (pthread_mutexattr_t *mutex_attr*, int
  protocol);

> Gets or sets the mutex protocol for dealing with priority inversions. Not implemented in RTAI Pthreads.
>
> *protocol* =   **PTHREAD_PRIO_NONE**
>   PTHREAD_PRIO_INHERIT
>   PTHREAD_PRIO_PROTECT
>
> **Errors**
>
> - **ENOSYS** – Priority scheduling is not supported
> - EINVAL – *attr* is not a mutex attribute object or *protocol* invalid (set only)
> - ENOTSUP – *protocol* value is not supported

Condition Variables

int pthread_cond_init (pthread_cond_t *cond*, const pthread_condattr_t
     **cond_attr*);

> Creates a condition variable object, *cond*. If non-null,
> *cond_attr* provides optional attributes.
>
> **Errors**
>
> - **EAGAIN** – insufficient resources other than memory
> - **ENOMEM** – insufficient memory for conditional
>   variable object
> - EBUSY – *cond* is already initialized
> - EINVAL – *cond_attr* is not a condition variable at-
>   tribute object

int pthread_cond_destroy (pthread_cond_t *cond*);

> Destroys an existing *cond* that is no longer needed.
>
> **Errors**
>
> - EBUSY – *cond* is in use
> - EINVAL – *cond* is not a condition variable

int pthread_cond_wait (pthread_cond_t *cond*, pthread_mutex_t *mutex*);

> Waits on a condition variable *cond* until awakened either by
> signal or broadcast. *mutex* is unlocked (before wait) and
> relocked (after wait) inside pthread_cond_wait().
>
> **Errors**
>
> - EINVAL – *cond* or *mutex* is not valid
> - EINVAL – different mutexes for concurrent waits
> - EINVAL – *mutex* is not owned by calling thread

int pthread_cond_timedwait (pthread_cond_t *cond*, pthread_mutex_t
     **mutex*, const struct timespec **time*);

> Waits on a condition variable *cond* until awakened either by
> signal or broadcast or until the absolute time *time* is reached.
> *mutex* is unlocked (before wait) and relocked (after wait)
> inside pthread_cond_wait().
>
> **Errors**
>
> - **ETIMEOUT** – *time* has expired
> - EINVAL – *cond, mutex* or *time* is not valid
> - EINVAL – different mutexes for concurrent waits
> - EINVAL – *mutex* is not owned by calling thread

int pthread_cond_signal (pthread_cond_t *cond*);

> Signals condition variable *cond* waking up one waiting thread.
> The order in which threads wake up depends on the schedul-
> ing policy.
>
> **Errors**
>
> - EINVAL – *cond* is not a condition variable

int pthread_cond_broadcast (pthread_cond_t *cond*);

> Like signal except that **all** waiting threads are awakened.
>
> **Errors**
>
> - EINVAL – *cond* is not a condition variable

Condition Variable Attributes

RTAI Pthreads does not support any condition variable attributes but it does
provide for creating and destroying an attribute object.

int pthread_condattr_init (pthread_condattr_t **cond_attr*);

> Initializes a condition variable attributes object with default
> values.

Errors

- **ENOMEM** – insufficient memory for attribute object

int pthread_condattr_destroy (pthread_condattr_t *cond_attr);

> Destroys a condition variable attribute object.
>
> Errors
>
> - EINVAL – *cond_attr* is not a condition variable attribute object

Message Queues

#include <rtai_pqueue.h>

The descriptions in this section do not distinguish between required and optional error codes. They list all of the error codes returned by the RTAI implementation.

mqd_t mq_open (char *mq name, int oflags, mode_t permissions, struct mq_attr *mq_attr);

> Creates a new message queue named *mq name* or opens an existing one for use by the calling thread. *oflags* controls the way the queue is accessed and, if necessary, opened as follows:
>
>> O_RDONLY, O_WRONLY or O_RDWR – normal access control
>>
>> O_NONBLOCK – don't block if the queue is full/empty
>>
>> O_CREAT – create the queue if it doesn't already exist.
>>
>> O_EXCL – when used with O_CREAT, return an error if the queue already exists
>
> *permissions* specifies the User/GroupOther, read/write/execute permissions for the queue. *mq_attr* specifies the "geometry" of the queue including:

>*mq_maxmsgs* – maximum number of messages the queue can hold
>
>*mq_msgsize* – maximum size of an individual message
>
>*mq_flags* – blocking, non-blocking behavior (only used by mq_setattr() and mq_getattr())
>
>*mq_curmsgs* – number of messages currently in the queue

permissions and *mq_attr* are only relevant if this call creates the message queue.

On success mq_open() returns a queue descriptor for use in subsequent message queue calls.

>**Errors**
>
>- ENOMEM – insufficient memory to create queue
>- EMFILE – no message queue descriptors available
>- EACCES – message queue exists and permissions in *oflags* denied or permission to create queue is denied.
>- EEXIST – message queue already exists and O_CREAT and O_EXCL were specified
>- EINVAL – *mq_name* or *mq_attr* invalid
>- ENOENT – message queue doesn't exist and O_CREAT not specified
>- ENAMETOOLONG – *mq_name* is too long

int mq_close (mqd_t *mq*);

>Closes the calling thread's link to message queue *mq*. The message queue still exists and any messages currently posted there may be accessed by other threads linked to *mq*.
>
>**Errors**
>
>- EBADF – *mq* is not a valid message queue identifier

int mq_unlink (char *mq_name);

> Destroys the message queue *mq_name* but only if no other threads have an open link to it. The queue's memory is deallocated and any messages remaining are lost. If other threads have open links to this queue it is marked for later deletion when the last open link is closed. Once mq_unlink() is called on a queue no further links can be opened to it.
>
> mq_unlink() returns 0 if it successfully destroys the queue. A positive return value is the number of links currently open to this queue.
>
> **Errors**
>
> - ENOENT   *mq_name* is not a valid message queue

int mq_send (mqd_t *mq*, const char *msg*, size_t *msglen*, unsigned int *prio*);

> Sends *msg* of length *msglen* and priority *prio* to message queue *mq*. Messages are placed on the queue in priority order. Within the same priority they are posted in FIFO order.
>
> **Errors**
>
> - EBADF – *mq* is not a valid message queue identifier
> - EINVAL – *prio* is greater than MQ_MAX_PRIO or the calling thread does not have proper queue access permissions.
> - EMSGSIZE – *msglen* is greater than *mq_msgsize* for this queue
> - EAGAIN – the queue is full and it is non-blocking

size_t mq_receive (mqd_t *mq*, char *msg_buff*, size_t *buflen*, unsigned int *prio*);

> Receives a message from queue *mq* and puts it in *msg_buff* of length *buflen*. The message priority is returned in *prio*. A positive return value is the length of the received message.

Errors

- EBADF – *mq* is not a valid message queue identifier
- EINVAL –calling thread does not have proper queue access permissions.
- EMSGSIZE – *buflen* is less than *mq_msgsize* for this queue
- EAGAIN – the queue is empty and it is non-blocking

int mq_notify (mqd_t *mq*, const struct sigevent **notify*);

        #include <asm/siginfo.h>   // struct sigevent

Allows the calling thread to arrange for asynchronous notification of the arrival of a message in *mq*. A message queue can only register one such notification request. A previously registered notification request can be removed by passing *notify* as NULL.

Errors

- EBADF – *mq* is not a valid message queue identifier
- -1 – notification request already registered with this queue or the request can't be cleared because it is owned by another thread.

int mq_getattr (mqd_t *mq*, struct mq_attr **mq_attr*);

Returns the attribute structure of message queue *mq*.

Errors

- EBADF – *mq* is not a valid message queue identifier

int mq_setattr (mqd_t *mq*, const struct mq_attr **new_attr*, struct mq_attr **old_attr*);

Sets the attributes of *mq* from *new_attr*. Only the *mq_flags* can be changed by this call. All others are unaffected. If *old_attr* is non-NULL, the original attribute structure of *mq* is stored here.

Errors

- EBADF – *mq* is not a valid message queue identifier
- EINVAL –calling thread does not have proper queue access permissions.

Why Software Should Not Have Owners

Richard Stallman, Free Software Foundation

Digital information technology contributes to the world by making it easier to copy and modify information. Computers promise to make this easier for all of us.

Not everyone wants it to be easier. The system of copyright gives software programs "owners", most of whom aim to withhold software's potential benefit from the rest of the public. They would like to be the only ones who can copy and modify the software that we use.

The copyright system grew up with printing—a technology for mass production copying. Copyright fit in well with this technology because it restricted only the mass producers of copies. It did not take freedom away from readers of books. An ordinary reader, who did not own a printing press, could copy books only with pen and ink, and few readers were sued for that.

Digital technology is more flexible than the printing press: when information has digital form, you can easily copy it to share it with others. This very flexibility makes a bad fit with a system like copyright. That's the reason for the increasingly nasty and draconian measures now used to enforce software copyright. Consider these four practices of the Software Publishers Association (SPA):

- Massive propaganda saying it is wrong to disobey the owners to help your friend.

- Solicitation for stool pigeons to inform on their coworkers and colleagues.

- Raids (with police help) on offices and schools, in which people are told they must prove they are innocent of illegal copying.

- Prosecution (by the US government, at the SPA's request) of people such as MIT's David LaMacchia, not for copying software (he is not accused of copying any), but merely for leaving copying facilities unguarded and failing to censor their use.

All four practices resemble those used in the former Soviet Union, where every copying machine had a guard to prevent forbidden copying, and where individuals had to copy information secretly and pass it from hand to hand as "samizdat." There is of course a difference: the motive for information control in the Soviet Union was political; in the US the motive is profit. But it is the actions that affect us, not the motive. Any attempt to block the sharing of information, no matter why, leads to the same methods and the same harshness.

Owners make several kinds of arguments for giving them the power to control how we use information:

- Name calling.

  Owners use smear words such as "piracy" and "theft," as well as expert terminology such as "intellectual property" and "damage," to suggest a certain line of thinking to the public—a simplistic analogy between programs and physical objects.

  Our ideas and intuitions about property for material objects are about whether it is right to *take an object away* from someone else. They don't directly apply to *making a copy* of something. But the owners ask us to apply them anyway.

- Exaggeration.

  Owners say that they suffer "harm" or "economic loss" when users copy programs themselves. But the copying has no direct effect on the owner, and it harms no one. The owner can lose only if the person who made the copy would otherwise have paid for one from the owner.

A little thought shows that most such people would not have bought copies. Yet the owners compute their "losses" as if each and every one would have bought a copy. That is exaggeration—to put it kindly.

- The law.

Owners often describe the current state of the law, and the harsh penalties they can threaten us with. Implicit in this approach is the suggestion that today's law reflects an unquestionable view of morality—yet at the same time, we are urged to regard these penalties as facts of nature that can't be blamed on anyone.

This line of persuasion isn't designed to stand up to critical thinking; it's intended to reinforce a habitual mental pathway.

It's elementary that laws don't decide right and wrong. Every American should know that, forty years ago, it was against the law in many states for a black person to sit in the front of a bus; but only racists would say sitting there was wrong.

- Natural rights.

Authors often claim a special connection with programs they have written, and go on to assert that, as a result, their desires and interests concerning the program simply outweigh those of anyone else—or even those of the whole rest of the world. (Typically companies, not authors, hold the copyrights on software, but we are expected to ignore this discrepancy.)

To those who propose this as an ethical axiom—the author is more important than you—I can only say that I, a notable software author myself, call it bunk.

But people in general are only likely to feel any sympathy with the natural rights claims for two reasons.

One reason is an overstretched analogy with material objects. When I cook spaghetti, I do object if someone else eats it, because then I cannot eat it. His action hurts me exactly as much as it benefits him;

only one of us can eat the spaghetti, so the question is, which? The smallest distinction between us is enough to tip the ethical balance.

But whether you run or change a program I wrote affects you directly and me only indirectly. Whether you give a copy to your friend affects you and your friend much more than it affects me. I shouldn't have the power to tell you not to do these things. No one should.

The second reason is that people have been told that natural rights for authors is the accepted and unquestioned tradition of our society.

As a matter of history, the opposite is true. The idea of natural rights of authors was proposed and decisively rejected when the US Constitution was drawn up. That's why the Constitution only *permits* a system of copyright and does not *require* one; that's why it says that copyright must be temporary. It also states that the purpose of copyright is to promote progress—not to reward authors. Copyright does reward authors somewhat, and publishers more, but that is intended as a means of modifying their behavior.

The real established tradition of our society is that copyright cuts into the natural rights of the public—and that this can only be justified for the public's sake.

■ Economics.

The final argument made for having owners of software is that this leads to production of more software.

Unlike the others, this argument at least takes a legitimate approach to the subject. It is based on a valid goal—satisfying the users of software. And it is empirically clear that people will produce more of something if they are well paid for doing so.

But the economic argument has a flaw: it is based on the assumption that the difference is only a matter of how much money we have to pay. It assumes that "production of software" is what we want, whether the software has owners or not.

People readily accept this assumption because it accords with our experiences with material objects. Consider a sandwich, for instance. You might well be able to get an equivalent sandwich either free or for a price. If so, the amount you pay is the only difference. Whether or not you have to buy it, the sandwich has the same taste, the same nutritional value, and in either case you can only eat it once. Whether you get the sandwich from an owner or not cannot directly affect anything but the amount of money you have afterwards.

This is true for any kind of material object—whether or not it has an owner does not directly affect what it *is*, or what you can do with it if you acquire it.

But if a program has an owner, this very much affects what it is, and what you can do with a copy if you buy one. The difference is not just a matter of money. The system of owners of software encourages software owners to produce something—but not what society really needs. And it causes intangible ethical pollution that affects us all.

What does society need? It needs information that is truly available to its citizens—for example, programs that people can read, fix, adapt, and improve, not just operate. But what software owners typically deliver is a black box that we can't study or change.

Society also needs freedom. When a program has an owner, the users lose freedom to control part of their own lives.

And above all society needs to encourage the spirit of voluntary cooperation in its citizens. When software owners tell us that helping our neighbors in a natural way is "piracy", they pollute our society's civic spirit.

This is why we say that free software is a matter of freedom, not price.

The economic argument for owners is erroneous, but the economic issue is real. Some people write useful software for the pleasure of writing it or for admiration and love; but if we want more software than those people write, we need to raise funds.

For ten years now, free software developers have tried various methods of finding funds, with some success. There's no need to make anyone rich; the median US family income, around $35k, proves to be enough incentive for many jobs that are less satisfying than programming.

For years, until a fellowship made it unnecessary, I made a living from custom enhancements of the free software I had written. Each enhancement was added to the standard released version and thus eventually became available to the general public. Clients paid me so that I would work on the enhancements they wanted, rather than on the features I would otherwise have considered highest priority.

The Free Software Foundation (FSF), a tax-exempt charity for free software development, raises funds by selling GNU CD-ROMs, T-shirts, manuals, and deluxe distributions, (all of which users are free to copy and change), as well as from donations. It now has a staff of five programmers, plus three employees who handle mail orders.

Some free software developers make money by selling support services. Cygnus Support, with around 50 employees [when this article was written], estimates that about 15 per cent of its staff activity is free software development—a respectable percentage for a software company.

Companies including Intel, Motorola, Texas Instruments and Analog Devices have combined to fund the continued development of the free GNU compiler for the language C. Meanwhile, the GNU compiler for the Ada language is being funded by the US Air Force, which believes this is the most cost-effective way to get a high quality compiler. [Air Force funding ended some time ago; the GNU Ada Compiler is now in service, and its maintenance is funded commercially.]

All these examples are small; the free software movement is still small, and still young. But the example of listener-supported radio in this country [the US] shows it's possible to support a large activity without forcing each user to pay.

As a computer user today, you may find yourself using a proprietary program. If your friend asks to make a copy, it would be wrong to refuse. Cooperation is more important than copyright. But underground, closet cooperation does not make for a good society. A person should aspire to live an upright life openly with pride, and this means saying "No" to proprietary software.

You deserve to be able to cooperate openly and freely with other people who use software. You deserve to be able to learn how the software works, and to teach your students with it. You deserve to be able to hire your favorite programmer to fix it when it breaks.

You deserve free software.

Updated: $Date: 2001/09/15 20:14:02 $ $Author: fsl $

Index

U

uCLinux, 103-104
user space, 11

W

web servers, embedded, 113-115
workstation, configuring, 55

Z

Z Shell (zsh), 29